NEUROSCIENCE RESEARCH PROGRESS

A CLOSER LOOK
AT NEUROTOXICITY

NEUROSCIENCE RESEARCH PROGRESS

Additional books and e-books in this series can be found on Nova's website under the Series tab.

NEUROSCIENCE RESEARCH PROGRESS

A CLOSER LOOK AT NEUROTOXICITY

GOKUL KRISHNA
EDITOR

Copyright © 2020 by Nova Science Publishers, Inc.

All rights reserved. No part of this book may be reproduced, stored in a retrieval system or transmitted in any form or by any means: electronic, electrostatic, magnetic, tape, mechanical photocopying, recording or otherwise without the written permission of the Publisher.

We have partnered with Copyright Clearance Center to make it easy for you to obtain permissions to reuse content from this publication. Simply navigate to this publication's page on Nova's website and locate the "Get Permission" button below the title description. This button is linked directly to the title's permission page on copyright.com. Alternatively, you can visit copyright.com and search by title, ISBN, or ISSN.

For further questions about using the service on copyright.com, please contact:
Copyright Clearance Center
Phone: +1-(978) 750-8400   Fax: +1-(978) 750-4470   E-mail: info@copyright.com.

NOTICE TO THE READER

The Publisher has taken reasonable care in the preparation of this book, but makes no expressed or implied warranty of any kind and assumes no responsibility for any errors or omissions. No liability is assumed for incidental or consequential damages in connection with or arising out of information contained in this book. The Publisher shall not be liable for any special, consequential, or exemplary damages resulting, in whole or in part, from the readers' use of, or reliance upon, this material. Any parts of this book based on government reports are so indicated and copyright is claimed for those parts to the extent applicable to compilations of such works.

Independent verification should be sought for any data, advice or recommendations contained in this book. In addition, no responsibility is assumed by the Publisher for any injury and/or damage to persons or property arising from any methods, products, instructions, ideas or otherwise contained in this publication.

This publication is designed to provide accurate and authoritative information with regard to the subject matter covered herein. It is sold with the clear understanding that the Publisher is not engaged in rendering legal or any other professional services. If legal or any other expert assistance is required, the services of a competent person should be sought. FROM A DECLARATION OF PARTICIPANTS JOINTLY ADOPTED BY A COMMITTEE OF THE AMERICAN BAR ASSOCIATION AND A COMMITTEE OF PUBLISHERS.

Additional color graphics may be available in the e-book version of this book.

Library of Congress Cataloging-in-Publication Data

ISBN: 978-1-53616-591-3
Library of Congress Control Number:2019953074

Published by Nova Science Publishers, Inc. † New York

CONTENTS

Preface		vii
Acknowledgments		xi
Chapter 1	Developmental Neurotoxicity: Insights into the Neurodevelopment and Behavior *Gokul Krishna*	1
Chapter 2	Extracellular Propagation of Lipid Raft Molecular Species Involved in Parkinson's Disease Neurodegeneration *Daniel Pereda, Ricardo Puertas-Avendaño, Milorad Zjalic, Marija Heffer, Ibrahim Gonzalez-Marrero, Miriam González-Gómez, Mario Diaz, and Raquel Marin*	53
Chapter 3	Neurotoxicity: Perspective in Age-Related Neurodegenerative Diseases *Shubhangini Tiwari and Sarika Singh*	85
About the Editor		127
Index		129
Related Nova Publications		139

PREFACE

There has been increased prevelance of neurotoxicity and neurodevelopmental toxicity owing to widespread release of industrial chemical compounds into the environment. A Closer Look at Neurotoxicity is a reference source that includes reviews, original research papers on current understanding and outcomes of neurotoxicant exposures on the brain of humans and experimental animals. In this book, the authors describe the mechanisms of environmental chemicals, produced substances and naturally occurring compounds, role of protein aggregates in promoting neurotoxicity, and neurotoxicity contribution to development of neurodevelopmental disorders and neurodegenerative diseases.

Chapter 1 provides a listing of individual environmental chemicals including pesticides, heavy metals and inorganic compounds, organic substances highlighting their effects, epidemiologic reports, cellular and molecular mechanisms of action, neurologic, behavioral, cognitive and neurodevelopmental disabilities from exposure to these neurotoxic agents. Mammalian neurodevelopment is a critical process and its interruption may have lifelong adverse consequence on brain structure and function, leading to neurodevelopmental disorders. Although the neurodevelopmental disorders are considered to be genetic origin, the ubiquitous exposure of human population to chemicals are of increased concern due to their ability to elicit developmental neurotoxicity. A widespread chemical exposure through diet, environment, occupation, consumer products or industrial setting during early life can have a profound impact on the brain morphology and developmental programming. A growing body of epidemiological and animal evidences suggest that neurotoxin exposure during a sensitive developmental stage can predispose to neurodevelopmental toxicities. Early-life chemical insult can produce progressive and cumulative neurotoxicity that might have delayed consequences leading to neurodegenerative diseases. The developmental neurotoxicants target specific events involving cell replication, differentiation, synaptic function and associated behavioral domains. Understanding the events in which the developing nervous system is vulnerable to chemical exposures has grown over period of years with increased prevalence of developmental neurotoxicity, developing effective interventions to counter the deleterious effects of these chemical candidates, holds great promise. The chapter illustrates current understanding of vast array of chemically diverse compounds and pathways implicated in developmental neurotoxicity.

Chapter 2, on extracellular vesicles that has been proposed act as signalosomes for the transfer of toxic proteins associated with neurodegenerative disease pathology. Several neurodegenerative diseases such as Alzheimer's and Parkinson's disease share a common mechanism involving aggregation of misfolded proteins which leads to progressive neuronal degeneration. The aggregates contain soluble protein or peptide that has been misfolded and accumulate early in the lifetime of the individual, but only manifest clinically in middle or late life. Cells release membranous vesicles known as exosomes involved in cell–cell signaling and transfer of prions. Lipids are essential components of exosomal membranes. Prions are the infectious particles that are responsible for transmissible neurodegenerative diseases. The chapter provides for role of the lipid rafts in the mechanisms of conformational transition and oligomerization of toxic protein markers but also in the cell-cell transmission of aberrant molecular species that may enhance the neuropathological progression. The results show that amyloidgenic protein (amyloid-β) and α-synuclein markers are abundantly secreted specifically in 30-90 nm size vesicles. Other protein raft markers known to participate in Alzheimer's disease (AD) pathology, such as flotillin-1 (a marker of lipid rafts) and the voltage dependent anion channel (VDAC) are also found in these exosomes. Noticeably, this 30-90 nm subclass shows a high content of ganglioside GM1 (a component of the sphingolipid signaling system known to affect neuronal proliferation), as an indicative of the potential involvement of this lipid class in toxic protein markers propagation.

Chapter 3 provides an outlook on the type of death and mechanisms involved in neurotoxicity which ultimately lead to neurodegenerative disorders. Neurodegenerative diseases are a heterogeneous group of diseases characterized by progressive and selective loss of anatomically related neuronal cells. As a result of increased life expectancy, neurodegenerative disorders are becoming more common. A variety of these diseases are characterized by shared mechanisms that contribute to cell death. Accumulating evidence points that oxidative and nitrosative stress, mitochondrial dysfunction, protofibril formation, the accumulation of misfolded proteins, and dysfunction of the ubiquitin-proteasome system are the primary events that commonly underlie the disease pathogenesis and contribute to ageing, which is the greatest risk factor for neurodegenerative diseases. Understanding the mechanisms and theories of pathogenesis in regulating neuronal death will serve as an avenue to help facilitate development of therapies focused on treatment and prevention of these diseases in the elderly population. The review provides an assessment of advances in neurotoxicity research on common themes occurring in several neurodegenerative diseases over the past decade.

ACKNOWLEDGMENTS

First, I would like to express my sincere gratitude to all the authors who contributed in the making of this book. Their support in addition to their own research projects, devoted a lot of time working on book chapters was excellent and I hope to see it continuously. Without the authors' expertise and contribution, finalizing this book would not have been feasible. The work of project manager Carra Feagaiga, in communicating with the editor, authors and working on the chapters was critical to the successful completion of the book. It is very much appreciated and very helpful in identifying key contributors who participated in the book publishing. A special thanks to President Nadya S. Columbus, Nova Science Publishers, Inc. for the support.

Gokul Krishna, PhD
Editor

In: A Closer Look at Neurotoxicity
Editor: Gokul Krishna
ISBN: 978-1-53616-591-3
© 2020 Nova Science Publishers, Inc.

Chapter 1

DEVELOPMENTAL NEUROTOXICITY: INSIGHTS INTO THE NEURODEVELOPMENT AND BEHAVIOR

Gokul Krishna[*], *PhD*
Department of Child Health,
University of Arizona College of Medicine-Phoenix,
Phoenix, AZ, US

ABSTRACT

Mammalian neurodevelopment is a critical process and its interruption may have lifelong adverse consequence on brain structure and function. Although several neurodevelopmental disorders are considered to be of genetic origin, the ubiquitous exposure of the human population exposure to chemicals such as pesticides, metals, alkenes and organic solvents are of increased concern due to their ability to elicit developmental neurotoxicity. Widespread chemical exposure through diet, environment, occupation, consumer products or industrial setting during early life can have a profound impact on the brain morphology and developmental programming. These include developmental delays,

---

[*] Corresponding Author's E-mail: gokulkrishna@email.arizona.edu; gokul2411@gmail.com

alterations in memory, cognition, hyperactive disorders and affective illness. Growing body of epidemiological and animal results suggest that neurotoxin exposure during a sensitive developmental stage can predispose to neurodevelopmental toxicity of chemical origin at levels below those required for systemic toxicity. Moreover, early-life chemical insult can produce progressive and cumulative neurotoxicity that might have delayed consequences leading to neurodegenerative diseases. The developmental neurotoxicants target specific events involving cell replication, differentiation, synaptic function and associated behavioral domains beginning during the prenatal maturation continuing into the postnatal period. Understanding the events in which the developing nervous system is vulnerable to chemical exposures has grown over a period of years involving mechanisms related to oxidative stress, mitochondrial dysfunction, interference in neurogenesis and thyroid function, synaptic failure, and neurotransmitter changes. With the increased prevalence of developmental neurotoxicity, developing effective interventions to counter the deleterious effects of these chemical candidates, holds great promise. Here, current understanding of pathways implicated in developmental neurotoxicity and models of developmental chemical exposure are reviewed that might help to provide some information to design effective treatments that prevent or ameliorate these harmful effects.

Keywords: developmental neurotoxicity, neurodevelopment, neurodegeneration, chemicals, prenatal, postnatal, behavior

ABBREVIATIONS

ASDs	Autism spectrum disorders
BDNF	Brain-derived neurotrophic factor
CREB	cyclic AMP response element-binding protein
DA	Dopamine
DNT	Developmental neurotoxicity
ERK	Extracellular-regulated kinase
GAP-43	Growth-associated protein-43
GD	Gestation day
GSK3β	Glycogen synthase kinase-3β

MPTP	1-methyl-4-phenyl-1,2,3,6-tetrahydropyridine
NMDA	*N*-methyl-*D*-aspartate
PCB	Polychlorinated Biphenyls
PD	Parkinson's disease
ROS	Reactive Oxygen Species
TH	Tyrosine hydroxylase
VSSC	Voltage-gated sodium channel

INTRODUCTION

The neurodevelopmental disorders include a group of disorders that affect critical periods of the ongoing central nervous system (CNS) maturation and development. The incidence of neurodevelopmental disorders including autism, attention deficit hyperactivity disorder and pediatric bipolar disorders are on the rise affecting various complex domains such as learning and memory, sensory and emotional functions, and behavior [1]. Developmental delays associated with neurotoxicity prolong children's capability to reach milestones in relation to cognitive or psychological skills including communication, social and motor behavior [2]. These disorders significantly affect the child's health with the potential for adverse effects in later life. Moreover, effective treatment of these conditions is extremely challenging since the deficits are often permanent resulting in a staggering increase in personal and societal costs.

Unifying theories have increasingly recognized that neurodevelopmental disorders are characterized by prototypical genetic factors [3]. However, in more than 50% of the cases, it is recognized that environmental factors underlie many neurodevelopmental diseases, with a genetic predisposition. Recent years has seen the increased concern that chemical exposures are contributing to the increasing incidence of neurodevelopmental toxicity in children [4].

This has provided important clues to understand the pathogenesis of the more common forms of the disease and period of vulnerability that occurs during development. The exquisite sensitive developmental stage is the controlled time periods during which environmental insult will have a deleterious effect. During this stage in humans, the immature brain will be susceptible to modifications and can be a major site of toxicity that can lead to permanent functional deficits [5]. The nervous system develops from the ectodermal cells of the embryo through a process called neurulation and develops into a complex network of interconnected cells.

On gestation day (GD) 18, the neural plate that faces amnion invaginates to form groove with folds on either side. The folds begin to form the neural tube and then separates from the overlying ectoderm. The cells detach from the ectoderm to form neural crest to generate sensory ganglia of spinal and cranial cells and Schwann cells in the peripheral nervous system. The neural tube later closes in the hindbrain area to create a caudal-to-rostral gradient for brain development [6]. Extensive research has been made assessing the *in utero* exposures to chemical neurotoxins on neurodevelopment. While epidemiologic data provides for much of the information, these studies are complemented by laboratory studies with animals models. The placenta protects the fetus against chemical toxins however, it does not serve as an effective barrier against chemical substances [7].

Although neurodegenerative diseases are mostly considered to be of genetic origin, it is increasingly recognized that genetic and environmental interactions play a crucial role. Particularly, environmental impact during the early developmental stage is important for the origin of neurodegenerative diseases. The delayed consequences of early-life insult lead to significant loss of neurons during development that has marked impact years or decades later as age advances [8, 9].

It is evident that multiple chemicals distributed in the environment can produce developmental neurotoxicity involving various mechanisms (Figure 1). The neurotoxic agents might also cause indirect toxicity to the developing brain by interactions with the genome. It is reported that gene-environment interaction causes a 25% of neurobehavioral disorders that involves epigenetic modification such as DNA methylation, histone modification or changes in non-protein-coding RNA [10].

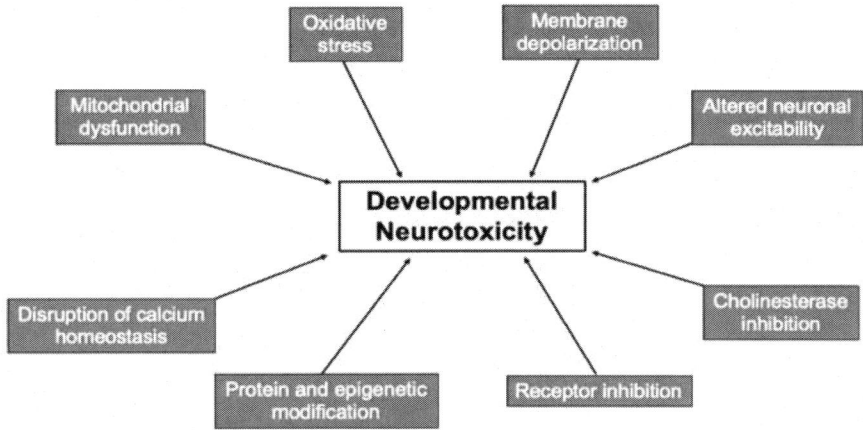

Figure 1. Potential mechanisms involved in developmental neurotoxicity.

1.1. ENHANCED FETAL AND INFANT SUSCEPTIBILITY TO NEUROTOXICITY

The developing brain of the fetus and infants are uniquely susceptible to neurotoxic exposures at impact levels that can have long term effects due to extended biologic retention time in the system [11]. The prenatal and postnatal period neurodevelopmental changes follow the formation of different brain systems and neuronal connections. These changes are essential for optimal neurodevelopment to enable proper development of the offspring. However, the vulnerability period during the complex, immature phase places the brain for higher risk of damage with a limited option for reversal and repair. The adverse effects often lead to permanent changes in the animal's physiology and behavior. Particularly, chemical exposures during developmental stages in infants and children are associated with several factors. The health hazards of chemicals are higher in children due to their frequent hand-to-mouth contact leading to increased permeability and higher absorption rates, and lower metabolism and excretion compared with the adults [12]. This is associated with many processes that occur during development over an extended period of time. The susceptibility depends on the time of exposure in relation to the brain developmental stage, the duration of exposure and the quantity of the exposed chemical. Moreover, increased brain exposure to toxic chemicals is due to incomplete formation of the blood-brain barrier that normally maintains homeostasis and protects the brain against harmful chemical substances [13].

The common possible sources of environmental toxin exposure in humans include air, water, food, and soil involving various routes such as inhalation, oral or skin. Variety of chemicals widely distributed in the environment including pesticides, metals, alkenes, organic solvents have been well studied for their developmental neurotoxicity (DNT). However, it is always not possible that chemical exposures occur in isolation but instead happens in combination with other risk factors, including other chemicals or toxicants. The multiple-hit theory for the influence of other risk factors when simultaneously target other critical areas of the brain by different mechanisms, the homeostasis balance would be severely dysregulated leading to cumulative damage [14].

1.2. Pesticides

Pesticides are commonly employed in agricultural practice and also in household settings to control insect and pests attack on crops and livestock. A variety of widely used pesticides in agriculture have known adverse neurologic effects [14]. Several epidemiologic and toxicological data report the pesticide exposure as a risk factor for Parkinson's disease (PD) [15-17]. Several epidemiologic studies and clinical reports have identified that exposures to certain synthetic chemicals are associated with an increased incidence of neurodegenerative diseases in later life. Although neurodegenerative diseases are considered to have a genetic basis, environmental toxicants are implicated as risk factors in PD. PD is the second most common neurodegenerative disease affecting about 1-2% of the worldwide population with primary hallmark involving degeneration of the nigrostriatal dopaminergic pathway, important for motor function. PD patients often present with clinical signs of tremor, rigidity and postural instability [18].

In addition to other mechanisms, oxidative stress, mitochondrial dysfunction, and impaired ubiquitin-proteasome system play a significant role in sporadic PD to cause dopaminergic loss [19]. The disease pathology accompanies the appearance of Lewy bodies, the cytoplasmic inclusions containing neurofilament proteins including α-synuclein and parkin. Several preclinical studies have employed pesticides as model neurotoxins to mimic characteristic features of PD. Although PD typically affects the older population, neurotoxic insult during early developmental stages might predispose to later-life neurodegeneration.

It is even proposed that deleterious events that occur early during development can have delayed consequences on the dopaminergic system following age-related neuronal loss or even cause permanent/ progressive neurotoxicity in the adults characteristic of PD. Moreover, autism spectrum disorders (ASDs) characterized by impaired social behaviors are increasing in several parts of the world which are linked to pesticide exposure with mechanisms involving oxidative stress, neuron excitability and immune dysfunction [20].

1.2.1. Pyrethroids

The pyrethroids are the most commonly used insecticides derived from pyrethrins (Chrysanthemum genus) [21] containing cyclopropane carboxylic acid moieties linked to alcohol through ester or ether bond. Common pyrethroids include type I - allethrin, bifenthrin, and type II - deltamethrin, cypermethrin, cyhalothrin, fenvalerate [22]. The voltage-gated sodium channel (VSSC) is critical to the normal function of the cells, for creating an inward sodium current that promotes action potential.

These VSSC consist of a subunit, which confers pore function and resembles other voltage-gated ion channels [23]. Upon exposure, pyrethroid insecticides primarily disturb VSSC function, delaying the opening of the VSSC and thus promoting a shift to more hyperpolarized membrane potential resulting in higher sodium ions to depolarize the neuronal membrane [24].

Type I compounds cause repeated firing of action potentials while type II compounds promote longer channel opening that leads to depolarization of membrane potential resulting in blockade of action potentials [25]. Several studies have shown adverse effects associated with pyrethroids on changes in VSSC function. Studies have even shown alterations in the nigrostriatal dopaminergic system. Cypermethrin exposure during postnatal days (PNDs) 5-19 significantly reduced the neurochemical levels (dopamine, 3,4-dihydro-xyphenylacetic acid) and tyrosine hydroxylase (TH) immunoreactivity in the striatum together with a reduction in motor behavior [26]. Table 1 provides a summary of the studies showing the developmental neurotoxicity of pyrethroids.

Table 1. Developmental neurotoxicity studies with pyrethroid insecticides in rodents

Compound	Dose and route of administration	Dosing duration	Outcome/Effect	Reference
Deltametrin	0.75 mg/kg, i.p.	GDs 7-10 and 11-14	Overexpression of reelin in the cells of the external granule cell layer (EGL), misalignment of the Purkinje cells, decreased spine density, dendritic length and width of the dendritic arbor.	[27]
Cypermethrin	5 and 20 mg/kg	Three times a week, from gestational day (GD) 6–7 to postnatal day 15 (PND 15)	Disturbed motor behavior, transcriptomic alterations in mitochondrial biogenesis, trafficking, and degradation of proteins.	[28]
	5 mg/kg, p.o.	GD 5 to 21	Alterations in histone acetylation and DNA methylation CYP1A- and 2B- isoenzymes, increase in the apoptotic signaling pathways.	[29]
Bifenthrin and β-cyfluthrin	100 µg/L	Gestation and lactation	Bifenthrin impaired pivoting in neonates, decreased open-field activity and rota-rod performance; enhanced brain oxidative stress. Cyfluthrin significantly impaired offspring growth.	[30]
Permethrin and cypermethrin	1.49 and 34.05 mg/kg p.o. respectively	PND 6 to PND 15	Increased spontaneous locomotor activity, reduced striatal levels of dopamine and homovanillic acid	[31]

1.2.2. Organophosphates

Organophosphate insecticides, commonly referred to as organophosphates (OPs), are primarily used in agriculture and as solvents in fuel additives, with higher rates of exposure among workers and residents of agricultural communities. Although the use of OPs has decreased, low-level occupational exposures still continue to be prevalent with global concerns on its neurotoxicity outcomes. The OPs that include – chlorpyrifos, diazinon, and parathion typically causes cholinesterase inhibition, cholinergic hyperstimulation and delayed (poly) neuropathy [32]. The oxon metabolites of the phosphorothionates are primarily responsible for the cholinergic neurotoxicity and disrupt processes involving neuronal differentiation and gliogenesis [33]. Although the organophosphates share a common mechanism of neurotoxicity, studies have revealed other neurotoxic effects attributable to oxidative stress [34-35], reduced expression of neurotrophic factors [36] and neural cell size [37]. The biotransformation of OPs involves microsomal enzymes of cytochrome P450 (CYP450) family to form phosphoxythiiran intermediates via oxidation of phosphorothionate followed by oxidative desulfuration to form oxon metabolite. This step is followed by metabolism by paraoxonase1 enzyme (PON1) inhibiting esterases [38].

1.2.3. Chlorpyrifos

Several potential developmental neurotoxicological studies with chlorpyrifos (CPF) have been conducted. CPF induces a disruption in neural cell maturation and synapse activity [39]. *In vivo* and *in vitro* studies with chlorpyrifos or diazinon has shown to target genes regulating the neuronal cell cycle and death [40]. Studies have shown divergent effects of diazinon and parathion on rat brain noradrenergic systems. Neonatal CPF exposure has shown to significantly reduce dopaminergic neurons with increased activation of microglia and astrocyte in the substantia nigra [41]. Microarray analysis showed that several genes associated with oxidative stress, mitochondrial dysfunction and brain development to be affected along with reduced acetylcholinesterase (AChE) activity in the cerebellum in response to repeated CPF exposure [42].

Several studies in rodent models have reported persistent changes in behavior in animals exposed to CPFs. The neurobehavioral consequences of chlorpyrifos exposure involve hyperkinesis[43], increased ultrasonic vocalization [44], lesser novelty exploration [45] and reduced spatial learning [46] and attention deficits [47]. Developmental defects associated with prenatal CPF exposure involves reduced birth weight [48] and delayed development of reflexes in mice [49]. A study by Vatanparast et al., [50] has shown that gestational exposure to CPF enhances while postnatal exposure reduces nicotinamide adenine dinucleotide phosphate (NADPH)-d(+)/nNOS-immunoreactive (IR) neurons within the basolateral nucleus of the amygdala without significant effect on passive avoidance response. In another study, GD 14-17 chlorpyrifos exposure induces sex-dimorphic behavioral effects on the offspring social behavior and enhances oxidative stress in a mouse model of idiopathic autism [51]. It is even shown that chlorpyrifos and cyfluthrin exposure causes immune toxicity associated with immunosuppression and neuroinflammation reducing antibody production and increased expression of proinflammatory cytokines [52, 53].

1.2.4. Diazinon and Parathion

Parathion exposure during PND 1-4 causes long-term behavioral alterations involving decreases tactile startle response [54]. The serotonergic system in the developing brain is among the most sensitive neurotransmitter systems to disruption by organophosphates contributing to developmental neurotoxicity. Early postnatal parathion administration to male rats caused cognitive impairments on radial arm maze accompanied by upregulation of 5HT receptors and enhancement in choline transporters (hemicholinium-3 binding) [55]. Studies have also described the distinct effects of neonatal exposures to parathion on serotonin and acetylcholine systems during adolescence and adulthood [56, 57]. Further, increased dopamine (DA) neurotransmitter turnover, disturbing synaptic activity has been previously observed with parathion [58]. Postnatal (PND 1-4) administration of diazinon to rats decreased while parathion caused an increase in norepinephrine levels [59]. Developmental exposures to diazinon affect indices of neural cell development (cell packing density and number) forming reactive gliosis in the prefrontal cortex and brainstem [60]. In vitro studies have revealed that diazinon enhanced tryptophan hydroxylase expression with a specific specific effect on synaptic function and behavior [61].

1.2.5. Paraquat

Paraquat (PQ, N,N-dimethyl-4-4-bipiridinium) continues to remain the most widely used herbicide. PQ shares its structural similarity to 1-methyl-4- phenylpyridinium (MPP^+), an active metabolite of 1-methyl-4-phenyl-1,2,3,6-tetrahydropyridine (MPTP) - a Parkinsonian agent [62]. PQ undergoes redox cycling with nicotinamide adenine dinucleotide phosphate-oxidase (NADPH) oxidase and nitric oxide synthase to yield monovalent cation [63]. Studies have identified Bak-dependent cell death mechanism involving cytochrome C release, caspase-3 and nuclear enzyme poly(ADP-ribose) (PARP) polymerase cleavage for PQ-induced neurotoxicity [64]. PQ exposure has shown to induce proteinaceous inclusions in the substantia nigra pars compacta, a neuropathological feature similar to observed in PD patients. PQ exposure has shown to selectively cause dopamine neuronal loss in the substantia nigra pars compacta and aggregation of α-synuclein [65]. The neurodevelopmental toxicity of PQ is attributed to its ability to induce oxidative stress to decrease cell proliferation and affect cell cycle processes [66]. PQ exposure in conjunction with other environmental chemical caused significant changes in the striatum in terms of neurochemical turnover and lower locomotor activity [67]. Gestational systemic PQ administration alters reproductive function in dams and elicits enhanced anxiety-like behavior in offspring [68]. Combined administration of PQ and maneb (a fungicide) from GD 6 until PND 21 impairs learning and memory associated with reduced expression of hippocampal cyclic AMP (cAMP) related genes, phosphorylation of the transcription factor cAMP response element-binding protein (CREB) and in turn lower brain-derived neurotrophic factor (BDNF) [69]. PQ treatment affects levels of amino acid transporters necessary for the development of cortical circuits and neurotransmission in the mouse parietal cortex with reduced motor activity among the pups [70, 71]. Hypoactive behavior and reduced dopamine levels were observed among offspring with neonatal exposure to PQ [72]. Moreover, *in vitro* studies in rat brain culture showed that PQ-induced adverse effects on dopaminergic, GABAergic and glutamatergic neurons with astrogliosis in the immature cultures [73].

1.2.6. Rotenone

Rotenone, an insecticide and used as a fish poison, belongs to the family of rotenoids, are the naturally occurring cytotoxic compounds extracted from the roots of *Lonchocarpus* plants [74]. Rotenone is highly lipophilic, readily crosses blood-brain barrier without specific transporter and accumulates in subcellular organelles (such as mitochondria). Rotenone reproduces features of PD with nigrostriatal dopaminergic degeneration and formation of cytoplasmic inclusions [75, 76]. It is known to potently inhibit complex I to impair oxidative phosphorylation with modest effects on ATP levels and cellular bioenergetics [77, 78]. Further, microglial activation derived reactive oxygen species (ROS) generation by rotenone has been found to impart neurotoxicity [79]. Rotenone neurotoxicity to immature ventral mesencephalic neurons and dopaminergic neurons has been shown to induce cell death and increased DA neuronal sensitivity respectively [80]. Studies in multicellular brain spheroids have indicated the developmental neurotoxicity during early cell differentiation stages together with higher ROS production, changes in pathways for Ca^{2+} reabsorption, synaptogenesis, peroxisome proliferator-activated receptors (PPAR) important for brain development [81]. Gestational rotenone administration reduced fetal weight, elevated levels of oxidative markers, depleted glutathione levels and perturbed activities of antioxidant enzymes in the maternal and fetal brain [82]. Childhood-onset neuropsychiatric disorder including attention deficit hyperactivity disorder (ADHD) is characterized by impaired attention, cognitive symptoms, hyperactive and impulsive behaviors. Rotenone oral administration to pups caused hyperactivity at juvenile and adulthood, and altered gene expression of molecules in control of cell cycle [83].

1.2.7. Polybrominated Diphenyl Ethers (PBDEs) and Polychlorinated Biphenyls (PCBs)

Organic substances have wide chemical application and are produced in a modern industrial setting as a result of various manufacturing processes are highly resistant to degradation. These organic compounds bioaccumulate, persists in the environment and are extremely hazardous to human health. Several studies suggest that these chemical substances affect neurodevelopment, reproduction and immune function [84]. The PBDEs and PCBs are persistent organic compounds that are lipophilic in nature. These substances are largely used as flame-retardants additives in electrics, plastics and in the manufacture of various electrical appliances [85]. The abundance of these chemicals poses a high risk, especially to toddlers and children.

Previous studies show developmental PBDEs exposure causing deficits in attention, psychomotor retardation, and enhanced impulsivity [86]. The dendritic spines are neuronal protrusions that receive input from excitatory synapse important for synaptic function and plasticity [87]. Morphological changes in the shape of dendritic spines are associated with many neurodevelopmental disorders [87]. PCB exposure in primary neuron-glia co-cultures showed increased axonal and dendritic growth in cortical and hippocampal pyramidal neurons [88].

The toxic potential of PBDEs involves alterations in cholinergic neurotransmission in the adult response to nicotine challenge [89]. Rats prenatally exposed to PBDE-99 adversely affects the cell signaling pathway enhancing the brain activity of the glutamate-nitric oxide-cGMP (3′,5′-cyclic guanosine monophosphate) pathway [90]. Neonatal PBDE 99 treatment significantly impairs motor function associated with the presence of its metabolites detected in the brain after 7 days after administration [91]. PCBs are the most common widely distributed environmental contaminant with several reports indicating their deleterious effects on human neurodevelopment and behavior [92-94]. The cytochrome P450 enzymes metabolize PCBs to hydroxylated PCBs [95].

Experimental studies have revealed that PCBs largely affects thyroid and estrogen function, [96, 97] and promotes ryanodine receptor activation [98]. Neonatal exposure to PCB congeners affected the activity of nicotinic receptors and cognitive performance in the adults [99]. It is even reported that PCB 152 and methylmercury co-exposure during neonatal period induces higher neurotoxic effects on spontaneous behavior and habituation in mice [100]. Gestational and lactational exposure to PCB 126 reduced weight gain and thyroxine (T4) levels [101]. Offspring perinatally exposed to PCB congeners displayed deficits in motor coordination with differentially expressed genes related to lipid metabolism and cell proliferation [102].

1.3. METALS AND INORGANIC COMPOUNDS

1.3.1. Lead

Lead (Pb), a highly toxic heavy metal primarily obtained from mining, manufacturing and burning fossil fuels. Although the use of Pb has been significantly reduced in the U.S., environmental and occupational low-level exposure continues in several countries with its use in fuels, paints and other substances. It has been noted that many children in U.S. present > 10 µg/dL of blood Pb levels, higher than the threshold concern defined by the Centers for Disease Control and Prevention (CDC). Pb exposure can affect various biological processes important for neurodevelopment causing a variety of behavioral syndromes including diminished school performance, antisocial personality and juvenile delinquency [103]. Numerous studies provide evidence for early-life Pb exposure and cognitive deficits in children [104, 105].

Table 2. Neurotoxic effects of Pb in animal models

Dose and route of administration of Pb	Dosing duration	Outcome/Effect	Reference
0, 0.5. 1.0 and 2 g/L in drinking water	GD 1 – PND 40	Reduced levels of PSD-95, nNOS and SYP	[122]
0.1% in drinking water	GD 1 – PND 21	Reduced levels of synaptic proteins including synaptotagmin-1, synaptophysin 25 and syntaxin-1 and PSD-95; lower BDNF levels	[123]
180 and 375 ppm in diet	30 days following weaning (PND 25–55)	Reduced mRNA expression levels of basic fibroblast growth factor, NMDA receptor subtypes, glutamate receptors and genes related to synaptic plasticity.	[124]
5 and 25 mg/L in drinking water	After mating until they gave birth to next generation	Hyperactivity behavior, increased levels of acetylated histone H3 and expresiion of p300	[125]

Pb neurotoxicity targets neuronal cells with evidence pointing to inhibition of *N*-methyl-*D*-aspartate (NMDA) receptors which play a critical role in synaptic transmission, neural network creation, and memory and learning functions [106]. Pb inhibits voltage-dependent anion channel expression, decreasing presynaptic neurotransmission, cellular energy metabolism by altering calcium buffering and synaptic plasticity important for learning and memory [107, 108]. Moreover, NMDA activity has been associated with neurotrophin (BDNF) signaling and it has been reported that Pb exposure reduced BDNF levels [109, 110] and CREB phosphorylation [111]. It is shown that Pb neurotoxicity mechanism during development involves hippocampal levels of mitogen-activated protein kinase (MAPK) signaling cascades, such as p38 and extracellular-regulated kinases (ERK)-1/2 [112]. Pb can alter epigenetic mechanisms affecting CNS maturation influencing the expression of markers for earlier stage progenitor cells and RNA-binding protein [113]. Dietary Pb administration has even shown to significantly affect DNA methyltransferases (DNMTs) 1 and 3 during developmental periods [114]. Studies in young rats have demonstrated that Pb reduces dentate granule cell neurogenesis altering the hippocampal cytoarchitecture [115]. Perinatal Pb exposure enhanced MDA levels [116], mitochondrial GSH levels along with reduction in activities of antioxidant enzymes – SOD1, GPx1 and GPx4 [117]. It has been reported that Pb affects oligodendrocytes and myelination by reducing levels of galactolipids in the developing rat brain [118]. Yu et al., demonstrated that Pb treatment from the first day of gestation until PND 40 impaired spatial memory of the offspring with reduced protein expression of D-amino acid oxidase AND NMDA subunits (NR1 and NR2A) [119]. Increased ROS generation and apoptotic markers (Bax/Bcl-2 ratio) were observed in offspring following prenatal and lactational lead acetate [$Pb(AC)_2$] exposure at 0.1 and 0.2% through drinking water [120]. Gestational Pb administration reduced glutamic acid decarboxylase (GAD67) and c-Kit expressions with a significant reduction in Purkinje cells in the cerebellar cortex [121].

1.3.2. Arsenic

Arsenic is a metalloid element present naturally in the environment and has long known to be used as a poison. It has been ranked as the hazardous substance under the section Comprehensive Environmental Response, Compensation and Liability Act by the Agency for Toxic Substances and Disease Registry (ATSDR) Priority List [126]. Arsenic presence has been detected worldwide in drinking water [127], mostly found in inorganic and organic forms. Arsenic exposure has been associated with a variety of adverse effects on humans including cancer has been a major concern. Several studies have confirmed the development of lung, bladder and skin cancer caused by inorganic arsenic ingestion [128-130]. Inorganic arsenic exposure is found to be particularly high in rice and rice products [131] due to significant arsenite mobilization through transport pathways [132].

Shorter escape latencies and deficits in spatial memory were observed among PND 42 offspring with a reduction in ERK and CREB expression in the hippocampus and cerebral cortex [133]. The impaired spatial learning has been linked to reduced expression of postsynaptic density protein-95 (PSD-95) and synaptophysin (SYP) along with alteration in the synaptic structure [134]. The developmental arsenic treatment has shown to reduce offspring reduced body weight and impaired cognition [135]. The corticosterone receptors in the hippocampus have a profound influence on the memory and hippocampus-dependent memories are influenced by glucocorticoid receptors [136]. Perinatal exposures have previously been reported to affect HPA functioning by enhancing the levels of corticosterone (CORT) while lowering glucocorticoid and mineralocorticoid receptors and reduced MAPK/ERK gene expression [137-140]. Hyperactivity of the HPA axis along with deficits in the negative feedback regulation of corticotrophin-releasing factor release may promote depressive illness [141]. Neuro-2a cells exposed to sodium arsenic revealed inhibited neurite outgrowth and genes encoding cytoskeletal components [142].

Reduced differentiated cell numbers and changes in neurogenesis-related genes regulating cell death, axonogenesis, neural differentiation and hippocampal morphology in mice after perinatal arsenic exposure [143]. Lactational transfer of arsenic has shown to significantly alter developmental landmarks with delayed incisor eruption and bilateral eye-opening with lower performance on slant board behavior in rats [144].

1.3.3. Methylmercury

Methylmercury (MeHg) is an organomercurial pollutant found in the aquatic environment. MgHgpoisoining became evident in the early 1950s in Minamata, Japan. Consumption of seafood by pregnant women resulted in fetal abnormalities and neurotoxicity characterized by blindness, microcephaly, mental and physical developmental retardations [145]. MeHg cytotoxicity involves several mechanisms that include perturbation of intracellular Ca^{2+} levels [146] and induction of oxidative stress by either overproduction of ROS [147].

Studies in animal models have demonstrated that the developing nervous system is susceptible to the neurotoxic effects of MeHG[148, 149]. The oxidative stress mechanism has been implicated in MeHg neurotoxicity targeting glutathione (GSH) [150]. Pregnant mice provided MeHg in drinking water caused inhibition of GSH along with the enzymes glutathione peroxidase (GSHPx) and glutathione reductase (GSHRd) in offspring brain [151]. In addition, MeHg-induced excitotoxicity with inhibition of cerebellar glutamate uptake has been reported in offspring [149]. Further evidence points to the role of Nrf2 and PI3K/Akt pathway mediating MeHg neurotoxicity [152]. The Nrf2 is the key sensor of oxidative stress in the regulation of gene expression regulating antioxidant proteins and xenobiotic detoxifying enzymes. MeHg has reported activating members of the mitogen-activated protein (MAPK) kinases, involving ERKs [153] and p38MAPK [154,

155]. A study in rats has shown developmental neurotoxicity in hippocampal downregulation of PI3K/Akt/mTOR pathway and activated/hypophosphorylated (Ser9) glycogen synthase kinase-3β (GSK3β) in the hippocampus following in utero MeHg treatment.

Table 3. Consequences of exposure to MeHg during the developmental period

Model	Dose and route	Duration	Effects on offspring	Reference
Rat	0.1, 0.4, 0.7, 1.0, 1.5, or 2.0mg/kg body weight/ day; gavage	GD 6 to PND 10	Delayed neural development, changes in GTPase signaling pathways	[156]
Mice	4 ppm in drinking water	Throughout gestation and PND 30	Decreased density of migrating cells, molecular layer widths and external granular layer width in the cerebellum.	[157]
Rats	1.5 mg/kg b.w.	GD 0 to GD 21	Hippocampal neurotoxicity characterized by mitochondrial dysfunction and Rho GTPase mRNA expression.	[158]
Mice	0.02mg/kg/day	Perinatal and weaning period	Behavioral deficits with increased locomotor function, motor impairment, and auditory dysfunction altered of lipid peroxidation, Na(+)/K(+)-ATPase activities, and nitric oxide.	[159]

1.3.4. Manganese

Manganese (Mn) is an essential trace element and has several beneficial effects on physiology essential for cellular functions [160]. Mn is readily available for absorption through the oral route and human exposures occur through drinking water contamination [161]. It is also known that airborne Mn is a major route of human exposure among welders and miners [162-164]. It has been reported that Mn absorption is 40% higher in neonates than adults [165]. The absorbed Mn usually binds to blood proteins (transferrin, alpha-2-macroglobulin) and is deposited mainly in the liver, bone and adipose tissue [166]. In children, the major route of exposure is ingestion [167] associated with cognitive deficits [168] and lower IQ scores [169, 170]. Pathways leading to Mn neurotoxicity involves oxidative stress [171, 172] in brain areas such as olfactory bulb and hypothalamus represent potentially sensitive areas [173]. It has been reported that mitochondrial Ca^{2+} uniporter sequesters and inhibits ATP synthesis, either the electron transport chain (complex II) or the glutamate/aspartate exchanger [174].

Studies in rats have shown that Mn exposure during post-natal days impairs motor coordination, balance and object recognition tasks [175]. Preweaning Mn overexposure caused decreased hippocampal CA1 long-term potentiation (LTP) and synucleinopathy [176]. Gestational exposure to manganese chloride ($MnCl_2$) from gestation day 10 to 21 showed apoptosis in subgranular zone, enhanced reelin-expressing and GABAergic interneurons showing aberration in neurogenesis [177]. Manganese chloride exposure caused the appearance of doublecortin positive cells in the dentate gyrus, indicating an increase of type-3 progenitor or immature granule cells [178]. It is also shown that Mn exposure to rats from PND 8-27 enhanced striatal p38 MAPK and Akt phosphorylation and cAMP-regulated phosphoprotein, 32k Da (DARPP-32). The study also revealed the enhanced activity of caspase and F_2-isoprostanes [179]. Reduced mRNA and protein expression of NMDA receptors - NR1, NR2A, and NR2B along with decreased levels of neurotrophic factors (CREB and BDNF) have been observed after 3 weeks of Mn administration [180].

It is known that Mn affects the basal ganglia characterized by neuronal loss, reactive gliosis and expression of inducible nitric oxide synthase (NOS2). Findings in mice reveal that Mn exposure caused increased activation of both microglia and astrocytes in the striatum

(St), globus pallidus (Gp), and substantia nigra pars reticulata (SNpr). Importantly, the study showed a higher expression of NOS2 in glia located in the Gp and SNpr. Additionally, a greater increase in the protein nitration (3-nitrotyrosine levels) was observed in dopamine- and DARPP-32 positive neurons of the striatum of preexposed as juveniles [181].

1.4. ORGANIC SUBSTANCES

1.4.1. Acrylamide

Acrylamide (ACR), a water-soluble unsaturated vinyl monomer is a type-2 alkene largely used in the manufacture of polymers materials, glues and plastics and in electrophoretic separation techniques. Higher industrial use of ACR has found its applicability in textile, ore processing and wastewater treatment [182]. Most commonly, acrylamide exposure among humans is through occupational setting, but however, it is also recognized that ACR is formed at very high temperatures above 120 °C during frying or baking variety of carbohydrate-rich foods [183]. The adverse health effects of ACR has been attributed to increased dietary intake of potato chips, fries, snacks, coffee and biscuits [184].

Long term ACR exposure affects both central and peripheral nervous systems causing skeletal muscle weakness, gait, ataxia, numbness, cognitive deficits and neuropathy [185, 186]. Studies in human and animal models have shown ACR induced loss of Purkinje cells with distal axon degeneration in the nervous system (reviewed by [187]) and even produces reproductive toxicity [188]. ACR genotoxicity is primarily due to its epoxide metabolite, glycidamide with high reactivity towards DNA [189] and several studies have documented ACR toxicity owing to irreversible covalent adducts formation with high nucleophilic cysteine sulfhydryl groups of neural proteins of homology-associated protein 1 (Keap1) leading to dissociation of nuclear factor erythroid 2-related factor 2 (Nrf2) [190]. Nitric oxide (NO) produced by amino acid L-arginine is involved in the regulation of neurotransmission and synaptic plasticity [191] and NO-sensitive proteome shares similarity with ACR-adduct proteome [192].

Population-based studies have revealed that dietary ACR intake crosses the placenta in humans to reduce birth weight and smaller head circumference [193, 194]. Increased oxidative stress has been demonstrated as one of the major mechanism involved in ACR neurotoxicity [195-198]. Gestational ACR exposure in rats markedly enhances markers of oxidative stress in maternal and fetal brain, reduction in DA levels, delayed cell proliferation in the cerebellum and decreased BDNF levels [199-203]. Perinatal ACR exposure alters motorneurons in brachial and lumbar spinal cord impairing motor reflexes in rats [201].

Prenatal and postnatal ACR administration induced behavioral deficits in offspring with impaired rotarod performance and reduced exploration on open field task [204, 205]. The growth-associated protein, GAP-43 is a marker of growth or plasticity is a key factor of neural regeneration [206] and synaptophysin regulates synaptic vesicle maturation involved in synaptogenesis [207]. Fetal ACR exposure inhibits GAP-43 and synaptophysin inhibiting neural proliferation and synaptic function with adverse effects on maternal behavior [208]. Transcriptomic analysis in cerebellum shows that ACR reduces expression of Nr4a2 gene involved in the development of dopaminergic neurons [209].

1.4.2. Toluene

Toluene is a volatile aromatic hydrocarbon used in several commercial products such as spray paints, glues, adhesives, and cleaning solutions. It has been reported that long-term abuse of toluene through inhalation causes cerebral ataxia and leukoencephalopathy [210-212]. Toulene neurotoxicity is mediated through inhibition of excitatory channels particularly the N-methyl-D-aspartate (NMDA) receptors [213, 214] and cholinergic nicotinic acetylcholine receptors [215] and causes a rapid increase in glutamate and taurine levels [216]. The acute neurotoxic effects of toluene include impairments in visual and cognition and behavioral deficits (reviewed by [217]). Early life toluene exposure enhances hippocampal caspase-3 immunoreactive cells and significantly impairs spatial learning mice [218] and neurogenesis [219]. Genomic analysis in rats shows acute toluene exposure to alter pathways associated with synaptic plasticity including synaptic length and mitochondrial function [220]. Cellular homeostasis regulating immune response and energy metabolism are significantly altered with acute toluene exposure promoting oxidative damage [221]. Age-related sensitivity of the nervous system to the neurotoxic insult of toluene has been demonstrated in rats [222].

SUMMARY AND FUTURE DIRECTIONS

This toxicologic evidence suggests that the neurotoxic actions caused by developmental neurotoxicants have emerged to be a significant risk factor for neurodevelopmental disorders. Moreover, it can also be possible that the influence genetic factors might play an important role that leads to neurodevelopmental disorders, or combined exposure to neurotoxins. Future studies prioritizing on detailed testing of harmful chemical agent(s) that are likely to cumulatively induce neurotoxicity to evaluate their effect on the developing system are highly needed. Further, regulatory mandates emphasizing for risk management and prevention are required to protect children from harmful effects of chemical exposure.

ACKNOWLEDGMENT

The author thanks Ms. Smitha Shekar for help with database searches and editing.

AUTHOR DISCLOSURE

None

REFERENCES

[1] Boivin, M. J., Kakooza, A. M., Warf, B. C., Davidson, L. L. (2015). Reducing neurodevelopmental disorders and disability through research and interventions. *Nature*, 527, S155.

[2] Shelton, J. F., Geraghty, E. M., Tancredi, D. J., Delwiche, L. D., Schmidt, E.J., Ritz, B., Hansen, R. L., Hertz-Picciotto, I. (2014). Neurodevelopmental disorders and prenatal residential proximity to agricultural pesticides: The CHARGE study. *Environ Health Perspect*, 122, 1103–1109.

[3] Sahin, M. and Sur, M. (2015). Genes, circuits, and precision therapies for autism and related neurodevelopmental disorders. *Science*, 350, aab3897.

[4] Grandjean, P. and Landrigan, P. J. (2006). Developmental neurotoxicity of industrial chemicals. *The Lancet*, 368, 2167–2178.

[5] Schettler, T. (2001). Toxic threats to neurologic development of children. *Environ Health Perspect*, 109, 813.

[6] Rice, D. and Barone, S. (2000). Critical periods of vulnerability for the developing nervous system: Evidence from humans and animal models. *Environ Health Perspect*, 108 (suppl 3),511–533.

[7] Andersen, H. R., Nielsen, J. B. and Grandjean, P. (2000). Toxicologic evidence of developmental neurotoxicity of environmental chemicals. *Toxicology*, 144, 121–127.

[8] Landrigan, P. J., Sonawane, B., Butler, R. N., Trasande, L., Callan, R. and Droller, D. (2005). Early environmental origins of neurodegenerative disease in later life. *Environ Health Perspect*, 113, 1230–1233.

[9] Miller, D. B. and O'Callaghan, J. P. (2008). Do early-life insults contribute to the late-life development of Parkinson and Alzheimer diseases? *Metabolism*, 57, S44–S49.

[10] Grafodatskaya, D., Chung, B., Szatmari, P. and Weksberg, R. (2010). Autism spectrum disorders and epigenetics. *J Am Acad Child Adolesc Psychiatry*, 49, 794–809.

[11] Goldman, L. R. and Koduru, S. (2000). Chemicals in the environment and developmental toxicity to children: A public health and policy perspective. *Environ Health Perspect*, 108, 443.

[12] Landrigan, P. J., Kimmel, C. A., Correa, A. and Eskenazi, B. (2004). Children's health and the environment: Public health issues and challenges for risk assessment. *Environ Health Perspect*, 112, 257.

[13] Moretti, R., Pansiot, J., Bettati, D., Strazielle, N., Ghersi-Egea, J. F., Damante, G., Fleiss, B., Titomanlio, L., Gressens., P. (2015). Blood-brain barrier dysfunction in disorders of the developing brain. *Front Neurosci*, 9, 40.

[14] Cory-Slechta, D. A., Thiruchelvam, M., Barlow, B. K. and Richfield, E. K. (2005). Developmental pesticide models of the Parkinson disease phenotype. *Environ Health Perspect*, 113, 1263–1270.

[15] De Lau, L. M. and Breteler, M. M. (2006). Epidemiology of Parkinson's disease. *Lancet Neurol*, 5, 525–535.

[16] Giasson, B. I. and Lee, V. M. Y. (2000). A new link between pesticides and Parkinson's disease. *Nat Neurosci*, 3, 1227.

[17] Brown, T. P., Rumsby, P. C., Capleton, A. C., Rushton L., Levy, L. S. (2005). Pesticides and Parkinson's disease—is there a link? *Environ Health Perspect*, 114, 156–164.

[18] Blesa, J. and Przedborski, S. (2014). Parkinson's disease: Animal models and dopaminergic cell vulnerability. *Front Neuroanat*, 8, 155.

[19] Moore, D. J., West, A. B., Dawson, V. L. and Dawson, T. M. (2005). Molecular pathophysiology of Parkinson's disease. *Annu Rev Neurosci*, 28, 57–87.

[20] Shelton, J. F., Hertz-Picciotto, I. and Pessah, I. N. (2012). Tipping the balance of autism risk: Potential mechanisms linking pesticides and autism. *Environ Health Perspect*, 120, 944–951.

[21] Casida, J. E. (1980). Pyrethrum flowers and pyrethroid insecticides. *Environ Health Perspect*, 34, 189.

[22] Wolansky, M. J. and Harrill, J. A. (2008). Neurobehavioral toxicology of pyrethroid insecticides in adult animals:a critical review. *Neurotoxicol Teratol*, 30, 55–78.

[23] Marban, E., Yamagishi, T. and Tomaselli, G. F. (1998). Structure and function of voltage-gated sodium channels. *J Physiol*, 508, 647–657.

[24] Narahashi, T. (1996) Neuronal ion channels as the target sites of insecticides. *Pharmacol Toxicol*, 79, 1–14.

[25] Ray, D. E. and Fry, J. R. (2006). A reassessment of the neurotoxicity of pyrethroid insecticides. *Pharmacol Ther*, 111, 174–193.

[26] Singh, A. K., Tiwari, M. N., Upadhyay, G., Patel, D. K., Singh, D. Prakash, O., Singh, M, P. Long term exposure to cypermethrin induces nigrostriatal dopaminergic neurodegeneration in adult rats: Postnatal exposure enhances the susceptibility during adulthood. *Neurobiol Aging*, 33, 404–415.

[27] Kumar, K., Patro, N. and Patro, I. (2013). Impaired structural and functional development of cerebellum following gestational

exposure of deltamethrin in rats: Role of reelin. *Cell Mol Neurobiol*, 33, 731–746.

[28] Laugeray, A., Herzine, A., Perche, O., Richard O., Montecot-Dubourg C., Menuet A., Mazaud-Guittot, S., Lesne, L., Jegou, B., Mortaud, S. (2017). In utero and lactational exposure to low-doses of the pyrethroid insecticide cypermethrin leads to neurodevelopmental defects in male mice-An ethological and transcriptomic study. *PloS One*, 12, e0184475.

[29] Singh, A., Agrahari, A., Singh, R., Yadav, S., Srivastava, V., Parmar, D. (2016). Imprinting of cerebral cytochrome P450s in offsprings prenatally exposed to cypermethrin augments toxicity on rechallenge. *Sci Rep*, 6, 37426.

[30] Syed, F., John, P. J. and Soni, I. (2016). Neurodevelopmental consequences of gestational and lactational exposure to pyrethroids in rats. *Environ Toxicol*, 31, 1761–1770.

[31] Nasuti, C., Gabbianelli, R., Falcioni, M. L., Di Stefano, A., Sozio, P., Cantalamessa, F. (2007). Dopaminergic system modulation, behavioral changes, and oxidative stress after neonatal administration of pyrethroids. *Toxicology*, 229, 194–205.

[32] Flaskos, J. (2014). The neuronal cytoskeleton as a potential target in the developmental neurotoxicity of organophosphorothionate insecticides. *Basic Clin Pharmacol Toxicol*, 115, 201–208.

[33] Flaskos, J. (2012). The developmental neurotoxicity of organophosphorus insecticides: A direct role for the oxon metabolites. *Toxicol Lett*, 209, 86–93.

[34] De Felice, A., Greco, A., Calamandrei, G. and Minghetti, L. (2016). Prenatal exposure to the organophosphate insecticide chlorpyrifos enhances brain oxidative stress and prostaglandin E 2 synthesis in a mouse model of idiopathic autism. *J Neuroinflammation*, 13, 149.

[35] López-Granero, C., Cañadas, F., Cardona, D., Yu, Y., Gimenez, E., Lazano, R., Avila, D. S., Aschner, M., Sanchez-Santed, F.

(2012). Chlorpyrifos-, diisopropylphosphorofluoridate-, and parathion-induced behavioral and oxidative stress effects: Are they mediated by analogous mechanisms of action? *Toxicol Sci,* 131, 206–216.

[36] Slotkin, T. A., Seidler, F. J. and Fumagalli, F. (2007). Exposure to organophosphates reduces the expression of neurotrophic factors in neonatal rat brain regions: Similarities and differences in the effects of chlorpyrifos and diazinon on the fibroblast growth factor superfamily. *Environ Health Perspect,* 115, 909.

[37] Qiao, D., Seidler, F. J., Tate, C. A., Cousins, M. M., Soltkin, T. A. (2003). Fetal chlorpyrifos exposure: Adverse effects on brain cell development and cholinergic biomarkers emerge postnatally and continue into adolescence and adulthood. *Environ Health Perspect,* 111, 536.

[38] Carr, R. L., Alugubelly, N. and Mohammed, A. N. (2018). Possible mechanisms of developmental neurotoxicity of organophosphate insecticides. *Adv Neurotoxicology, 2,* 145–188.

[39] Mauro, R. E. and Zhang, L. (2007). Unique insights into the actions of CNS agents:lessons from studies of chlorpyrifos and other common pesticides. *Cent Nerv Syst Agents Med Chem,* 7, 183–199.

[40] Slotkin, T. A. and Seidler, F. J. (2012). Developmental neurotoxicity of organophosphates targets cell cycle and apoptosis, revealed by transcriptional profiles *in vivo* and *in vitro*. *Neurotoxicol Teratol,* 34, 232–241.

[41] Zhang, J., Dai, H., Deng, Y., Tian, J., Zhang, C., Hu, Z., Bing, G., Zhao, L. (2015). Neonatal chlorpyrifos exposure induces loss of dopaminergic neurons in young adult rats. *Toxicology,* 336, 17–25.

[42] Cole, T. B., Beyer, R. P., Bammler, T. K., Park, S. S., Farin, F. M., Costa, L. G., Furlong, C. E. (2011). Repeated developmental exposure of mice to chlorpyrifos oxon is associated with

paraoxonase 1 (PON1)-modulated effects on cerebellar gene expression. *Toxicol Sci*, 123, 155–169.

[43] Cole, T. B., Fisher, J. C., Burbacher, T. M., Costa, L. G., Furlong, C. E. (2012). Neurobehavioral assessment of mice following repeated postnatal exposure to chlorpyrifos-oxon. *Neurotoxicol Teratol*, 34, 311–322.

[44] Venerosi, A., Ricceri, L., Scattoni, M. L. and Calamandrei, G. (2009). Prenatal chlorpyrifos exposure alters motor behavior and ultrasonic vocalization in CD-1 mouse pups. *Environ Health*, 8, 12.

[45] Laporte, B., Gay-Quéheillard, J., Bach, V. and Villégier, A. S. (2018). Developmental neurotoxicity in the progeny after maternal gavage with chlorpyrifos. *Food Chem Toxicol*, 113, 66–72.

[46] López-Granero, C., Cardona, D., Giménez, E., Lozano, R., Barril, J., Sanchez-Santed, F., Canadas, F. (2013). Chronic dietary exposure to chlorpyrifos causes behavioral impairments, low activity of brain membrane-bound acetylcholinesterase, and increased brain acetylcholinesterase-R mRNA. *Toxicology*, 308, 41–49.

[47] Marks, A. R., Harley, K., Bradman, A., Kogut, K., Barr, D. B., Johnson, C., Calderon, N., Eskenazi, B. (2010). Organophosphate pesticide exposure and attention in young Mexican-American children: the CHAMACOS study. *Environ Health Perspect*, 118, 1768.

[48] Whyatt, R. M., Rauh, V., Barr, D. B., Camann, D. E., Andrews, h. F., Garfinkel, R., Hoepner, L. A., Diaz, D., Dietrich, J., Reyes, A., Tang, D. (2004). Prenatal insecticide exposures and birth weight and length among an urban minority cohort. *Environ Health Perspect*, 112, 1125.

[49] Lan, A., Kalimian, M., Amram, B. and Kofman, O. (2017). Prenatal chlorpyrifos leads to autism-like deficits in C57Bl6/J mice. *Environ Health*, 16, 43.

[50] Vatanparast, J., Naseh, M., Baniasadi, M. and Haghdoost-Yazdi, H. (2013). Developmental exposure to chlorpyrifos and diazinon differentially affect passive avoidance performance and nitric oxide synthase-containing neurons in the basolateral complex of the amygdala. *Brain Res*, 1494, 17–27.

[51] De Felice, A., Greco, A., Calamandrei, G. and Minghetti, L. (2016). Prenatal exposure to the organophosphate insecticide chlorpyrifos enhances brain oxidative stress and prostaglandin E2 synthesis in a mouse model of idiopathic autism. *J Neuroinflammation*, 13, 149.

[52] Rooney, A. A., Matulka, R. A. and Luebke, R. W. (2003). Developmental atrazine exposure suppresses immune function in male, but not female Sprague-Dawley rats. *Toxicol Sci*, 76, 366–375.

[53] Mense, S. M., Sengupta, A., Lan, C., Zhou, M., Bentsman, G., Volsky, D. J., Whyatt, R. M., Perera, F. P., Zhang, L. (2006). The common insecticides cyfluthrin and chlorpyrifos alter the expression of a subset of genes with diverse functions in primary human astrocytes. *Toxicol Sci*, 93, 125–135.

[54] Timofeeva, O. A., Sanders, D., Seemann, K., Yang, L., Hermanson, D., Regenbogen, S., Agoos, S., Kallepalli, A., Rastogi., A., Braddy, D., Wells, C. (2008). Persistent behavioral alterations in rats neonatally exposed to low doses of the organophosphate pesticide, parathion. *Brain Res Bull*, 77, 404–411.

[55] Levin, E. D., Timofeeva, O. A., Yang, L., Petro, A., Ryde, I. T., Wrench, N., Seidler, F. J., Slotkin., T. A. (2010). Early postnatal parathion exposure in rats causes sex-selective cognitive impairment and neurotransmitter defects which emerge in aging. *Behav Brain Res*, 208, 319–327.

[56] Slotkin, T. A., Bodwell, B. E., Ryde, I. T., Levin, E. D., Seidler, F. J. (2008). Exposure of neonatal rats to parathion elicits sex-selective impairment of acetylcholine systems in brain regions

during adolescence and adulthood. *Environ Health Perspect*, 116, 1308.

[57] Slotkin, T. A., Levin, E. D. and Seidler, F. J. (2009). Developmental neurotoxicity of parathion: Progressive effects on serotonergic systems in adolescence and adulthood. *Neurotoxicol Teratol*, 31, 11–17.

[58] Slotkin, T. A., Wrench, N., Ryde, I. T., Lassiter, T. L., Levin, E. D. Seidler, F. J. (2009). Neonatal parathion exposure disrupts serotonin and dopamine synaptic function in rat brain regions:modulation by a high-fat diet in adulthood. *Neurotoxicol Teratol*, 31, 390–399.

[59] Slotkin, T. A., Skavicus, S. and Seidler, F. J. (2017). Diazinon and parathion diverge in their effects on development of noradrenergic systems. *Brain Res Bull*, 130, 268–273.

[60] Slotkin, T. A., Bodwell, B. E., Levin, E. D. and Seidler, F. J. (2007). Neonatal exposure to low doses of diazinon: Long-term effects on neural cell development and acetylcholine systems. *Environ Health Perspect*, 116, 340–348.

[61] Slotkin, T. A. and Seidler, F. J. (2008). Developmental neurotoxicants target neurodifferentiation into the serotonin phenotype:chlorpyrifos, diazinon, dieldrin and divalent nickel. *Toxicol Appl Pharmacol*, 233, 211–219.

[62] Langston, J. W., Ballard, P., Tetrud, J. W. and Irwin, I. (1983). Chronic Parkinsonism in humans due to a product of meperidine-analog synthesis. *Science*, 219, 979–980.

[63] Rappold, P. M., Cui, M., Chesser, A. S., Tibbett, J., Grima, J. C., Duan, L., Sen, N., Javitch., J. A., Tieu, K. (2011). Paraquat neurotoxicity is mediated by the dopamine transporter and organic cation transporter-3. *Proc Natl Acad Sci*, 108, 20766–20771.

[64] Fei, Q., McCormack, A. L., Di Monte, D. A. and Ethell, D. W. (2008). Paraquat neurotoxicity is mediated by a Bak-dependent mechanism. *J Biol Chem*, 283, 3357–3364.

[65] Manning-Bog, A. B., McCormack, A. L., Li, J., Uversky, V. N., Fink, A. L., Di Monte, D. A. (2002). The herbicide paraquat causes up-regulation and aggregation of α-synuclein in mice paraquat and α-synuclein. *J Biol Chem*, 277, 1641–1644.

[66] Colle, D., Farina, M., Ceccatelli, S. and Raciti, M. (2018). Paraquat and maneb exposure alters rat neural stem cell proliferation by inducing oxidative stress: New insights on pesticide-induced neurodevelopmental toxicity. *Neurotox Res*, 1–14.

[67] Barlow, B. K., Richfield, E. K., Cory-Slechta, D. A. and Thiruchelvam, M. (2004). A fetal risk factor for Parkinson's disease. *Dev Neurosci*, 26, 11–23.

[68] Ait-Bali, Y., Ba-M'hamed, S. and Bennis, M. (2016). Prenatal Paraquat exposure induces neurobehavioral and cognitive changes in mice offspring. *Environ Toxicol Pharmacol*, 48, 53–62.

[69] Li, B., He, X., Sun, Y. and Li, B. (2016). Developmental exposure to paraquat and maneb can impair cognition, learning and memory in Sprague-Dawley rats. *Mol Biosyst*, 12, 3088–3097.

[70] Benitez-Diaz, P. and Miranda-Contreras, L. (2009). Effects of prenatal exposure to paraquat on the development of amino acid synaptic transmission in mouse cerebral parietal cortex. *Invest Clin*, 50, 465–478.

[71] Miranda-Contreras, L., Dávila-Ovalles, R., Benítez-Díaz, P., Peña-Contreras, Z. and Palacios-Prü, E. (2005). Effects of prenatal paraquat and mancozeb exposure on amino acid synaptic transmission in developing mouse cerebellar cortex. *Dev Brain Res*, 160, 19–27.

[72] Fredriksson, A., Fredriksson, M. and Eriksson, P. (1993). Neonatal exposure to paraquat or MPTP induces permanent changes in striatum dopamine and behavior in adult mice. *Toxicol Appl Pharmacol*, 122, 258–264.

[73] Sandström, J., Broyer, A., Zoia, D., Schilt, C., Greggio, C. (2017). Potential mechanisms of development-dependent adverse effects of the herbicide paraquat in 3D rat brain cell cultures. *Neurotoxicology*, 60, 116–124.

[74] Uversky, V. N. (2004). Neurotoxicant-induced animal models of Parkinson's disease: Understanding the role of rotenone, maneb and paraquat in neurodegeneration. *Cell Tissue Res*, 318, 225–241.

[75] Betarbet, R., Sherer, T. B., MacKenzie, G., Garcia-Osuma, M., Panov, A. V., Greenamyre, J. T. (2000). Chronic systemic pesticide exposure reproduces features of Parkinson's disease. *Nat Neurosci*, 3, 1301.

[76] Sherer, T. B., Kim, J. H., Betarbet, R. and Greenamyre, J. T. (2003). Subcutaneous rotenone exposure causes highly selective dopaminergic degeneration and α-synuclein aggregation. *Exp Neurol*, 179, 9–16.

[77] Tieu, K. (2011). A guide to neurotoxic animal models of Parkinson's disease. *Cold Spring Harb Perspect Med*, 1, a009316.

[78] Sherer, T. B., Betarbet, R., Testa, C. M., Seo, B. B., Richardson, J. R., Kim, J. H., Miller, G.W., Yagi, T., Matsuno-Yagi, A. and Greenamyre, J. T. (2003). Mechanism of toxicity in rotenone models of Parkinson's disease. *J Neurosci*, 23, 10756–10764.

[79] Gao, H. M., Hong, J. S., Zhang, W. and Liu, B. (2002). Distinct role for microglia in rotenone-induced degeneration of dopaminergic neurons. *J Neurosci*, 22, 782–790.

[80] Bollimpelli, V. S. and Kondapi, A. K. (2015). Differential sensitivity of immature and mature ventral mesencephalic neurons to rotenone induced neurotoxicity *in vitro*. *Toxicol In Vitro*, 30, 545–551.

[81] Pamies, D., Block, K., Lau, P., Gribaldo, L., Pardo, C. A., Barreras, P., Smirnova, L., Wiersma, D., Zhao, L., Harris, G., Hartung, T. (2018). Rotenone exerts developmental neurotoxicity

in a human brain spheroid model. *Toxicol Appl Pharmacol*, 354, 101–114.

[82] Krishna, G. and Muralidhara. (2018). Oral supplements of inulin during gestation offsets rotenone-induced oxidative impairments and neurotoxicity in maternal and prenatal rat brain. *Biomed Pharmacother*, 104, 751–762.

[83] Ishido, M., Suzuki, J. and Masuo, Y. (2017). Neonatal rotenone lesions cause onset of hyperactivity during juvenile and adulthood in the rat. *Toxicol Lett*, 266, 42–48.

[84] Damstra, T. (2002). Potential effects of certain persistent organic pollutants and endocrine disrupting chemicals on the health of children. *J Toxicol Clin Toxicol*, 40, 457–465.

[85] Winneke, G. (2011). Developmental aspects of environmental neurotoxicology:lessons from lead and polychlorinated biphenyls. *J Neurol Sci*, 308, 9–15.

[86] Berghuis, S. A., Bos, A. F., Sauer, P. J. and Roze, E. (2015). Developmental neurotoxicity of persistent organic pollutants:an update on childhood outcome. *Arch Toxicol*, 89, 687–709.

[87] Nimchinsky, E. A., Sabatini, B. L. and Svoboda, K. (2002). Structure and function of dendritic spines. *Annu Rev Physiol*, 64, 313–353.

[88] Sethi, S., Keil, K. P., Chen, H., Hayakawa, K., Li, X., Lin, Y., Lehmler, H. J., Puschner, B. and Lein, P. J. (2017). Detection of 3, 3′-dichlorobiphenyl in human maternal plasma and its effects on axonal and dendritic growth in primary rat neurons. *Toxicol Sci*, 158, 401–411.

[89] Viberg, H., Fredriksson, A. and Eriksson, P. (2002). Neonatal exposure to the brominated flame retardant 2, 2, 4, 4, 5-pentabromodiphenyl ether causes altered susceptibility in the cholinergic transmitter system in the adult mouse. *Toxicol Sci*, 67, 104–107.

[90] Llansola, M., Erceg, S., Monfort, P., Montoliu, C., Felipo, V. (2007). Prenatal exposure to polybrominateddiphenylether 99

enhances the function of the glutamate? nitric oxide? cGMP pathway in brain *in-vivo* and in cultured neurons. *Eur J Neurosci*, 25, 373–379.

[91] Eriksson, P., Viberg, H., Jakobsson, E., Orn, U., Fredriksson, A. (2002). A brominated flame retardant, 2, 2, 4, 4, 5-pentabromodiphenyl ether: Uptake, retention, and induction of neurobehavioral alterations in mice during a critical phase of neonatal brain development. *Toxicol Sci*, 67, 98–103.

[92] Stewart, P., Reihman, J., Lonky, E., Darvill, T., Pagano., J. (2000). Prenatal PCB exposure and neonatal behavioral assessment scale (NBAS) performance. *Neurotoxicol Teratol*, 22, 21–29.

[93] Darvill, T., Lonky, E., Reihman, J., Stewart, P., Pagano, J. (2000). Prenatal exposure to PCBs and infant performance on the Fagan Test of Infant Intelligence. *Neurotoxicology*, 21, 1029–1038.

[94] Stewart, P., Fitzgerald, S., Reihman, J., Gump, B., Lonky, E., Darvill, T., Pagano, J., Hauser, P. (2003). Prenatal PCB exposure, the corpus callosum, and response inhibition. *Environ Health Perspect*, 111, 1670.

[95] Grimm, F. A., Hu, D., Kania-Korwel, I., Lehmler, H. J., Ludewing, G., Hornbuckle, K. C., Duffel, M. W., Bergman, A., Robertson, L. W. (2015). Metabolism and metabolites of polychlorinated biphenyls. *Crit Rev Toxicol*, 45, 245–272.

[96] Zoeller, R. T., Dowling, A. L. and Vas, A. A. (2000). Developmental exposure to polychlorinated biphenyls exerts thyroid hormone-like effects on the expression of RC3/neurogranin and myelin basic protein messenger ribonucleic acids in the developing rat brain. *Endocrinology*, 141, 181–189.

[97] Steinberg, R. M., Juenger, T. E. and Gore, A. C. (2007). The effects of prenatal PCBs on adult female paced mating reproductive behaviors in rats. *Horm Behav*, 51, 364–372.

[98] Pessah, I. N., Cherednichenko, G. and Lein, P. J. (2010). Minding the calcium store: Ryanodine receptor activation as a convergent mechanism of PCB toxicity. *Pharmacol Ther*, 125, 260–285.

[99] Eriksson, P. and Fredriksson, A. (1996). Developmental neurotoxicity of four ortho-substituted polychlorinated biphenyls in the neonatal mouse. *Environ Toxicol Pharmacol*, 1, 155–165.

[100] Fischer, C., Fredriksson, A. and Eriksson, P. (2008). Neonatal co-exposure to low doses of an ortho-PCB (PCB 153) and methyl mercury exacerbate defective developmental neurobehavior in mice. *Toxicology*, 244, 157–165.

[101] Rice, D. C. and Hayward, S. (1999). Effects of exposure to 3, 3', 4, 4', 5-pentachlorobiphenyl (PCB 126) throughout gestation and lactation on behavior (concurrent random interval–random interval and progressive ratio performance) in rats. *Neurotoxicol Teratol*, 21, 679–687.

[102] De Boever, P., Wens, B., Boix, J., Felipo, V., Schoeters, G. (2013). Perinatal exposure to purity-controlled polychlorinated biphenyl 52, 138, or 180 alters toxicogenomic profiles in peripheral blood of rats after 4 months. *Chem Res Toxicol*, 26, 1159–1167.

[103] Dietrich, K. N., Douglas, R. M., Succop, P. A., Berger, O. G., Bornschein, R. L. (2001). Early exposure to lead and juvenile delinquency. *Neurotoxicol Teratol*, 23, 511–518.

[104] Lanphear, B. P., Hornung, R., Khoury, J., Yolton, K., Baghurst, P., Bellinger, D. C. Canfield, R. L., Dietrich, K. N., Bornschein, R., Greene, T., Rothenberg, S. J. (2005). Low-level environmental lead exposure and children's intellectual function:an international pooled analysis. *Environ Health Perspect*, 113, 894.

[105] Lanphear, B. P., Dietrich, K., Auinger, P. and Cox, C. (2000). Cognitive deficits associated with blood lead concentrations < 10 microg/dL in US children and adolescents. *Public Health Rep*, 115, 521.

[106] Marchetti, C. (2003). Molecular targets of lead in brain neurotoxicity. *Neurotox Res*, 5, 221–235.

[107] Prins, J. M., Park, S. and Lurie, D. I. (2009). Decreased expression of the voltage-dependent anion channel in differentiated PC-12 and SH-SY5Y cells following low-level Pb exposure. *Toxicol Sci*, 113, 169–176.

[108] Prins, J. M., Brooks, D. M., Thompson, C. M. and Lurie, D. I. (2010). Chronic low-level Pb exposure during development decreases the expression of the voltage-dependent anion channel in auditory neurons of the brainstem. *Neurotoxicology*, 31, 662–673.

[109] Neal, A. P., Stansfield, K. H., Worley, P. F., Thompson, R. E., Guilarte, T. R. (2010). Lead exposure during synaptogenesis alters vesicular proteins and impairs vesicular release: potential role of NMDA receptor–dependent BDNF signaling. *Toxicol Sci*, 116, 249–263.

[110] Neal, A. P. and Guilarte, T. R. (2010). Molecular neurobiology of lead (Pb^{2+}): Effects on synaptic function. *Mol Neurobiol*, 42, 151–160.

[111] Toscano, C. D., Hashemzadeh-Gargari, H., McGlothan, J. L. and Guilarte, T. R. (2002). Developmental Pb2+ exposure alters NMDAR subtypes and reduces CREB phosphorylation in the rat brain. *Dev Brain Res*, 139, 217–226.

[112] Cordova, F. M., Rodrigues, A. L. S., Giacomelli, M. B., Oliveira, C. S., Posser, T., Dunkley, P. R., Leal, R. B. (2004). Lead stimulates ERK1/2 and p38MAPK phosphorylation in the hippocampus of immature rats. *Brain Res*, 998, 65–72.

[113] Senut, M. C., Sen, A., Cingolani, P., Shaik, A., Land, S. J., Ruden, D. M. (2014). Lead exposure disrupts global DNA methylation in human embryonic stem cells and alters their neuronal differentiation. *Toxicol Sci*, 139, 142–161.

[114] Schneider, J. S., Kidd, S. K. and Anderson, D. W. (2013). Influence of developmental lead exposure on expression of DNA

methyltransferases and methyl cytosine-binding proteins in hippocampus. *Toxicol Lett*, 217, 75–81.

[115] Verina, T., Rohde, C. A. and Guilarte, T. R. (2007). Environmental lead exposure during early life alters granule cell neurogenesis and morphology in the hippocampus of young adult rats. *Neuroscience*, 145, 1037–1047.

[116] Antonio-García, M. T. and Massó-Gonzalez, E. L. (2008). Toxic effects of perinatal lead exposure on the brain of rats: Involvement of oxidative stress and the beneficial role of antioxidants. *Food Chem Toxicol*, 46, 2089–2095.

[117] Baranowska-Bosiacka, I., Gutowska, I., Marchlewicz, M., Marchetti, C., Kurzawski, M., Dziedziejko, V, Kolasa, A., Olszewska, M., Rybicka, M., Safranow, K., Nowacki, P. (2012). Disrupted pro-and antioxidative balance as a mechanism of neurotoxicity induced by perinatal exposure to lead. *Brain Res*, 1435, 56–71.

[118] Deng, W. and Poretz, R. D. (2001). Lead exposure affects levels of galactolipid metabolic enzymes in the developing rat brain. *Toxicol Appl Pharmacol*, 172, 98–107.

[119] Yu, H., Li, T., Cui, Y., Liao, Y, Wang, G., Gao, L., Zhao, F., Jun, Y. (2014). Effects of lead exposure on D-serine metabolism in the hippocampus of mice at the early developmental stages. *Toxicology*, 325, 189–199.

[120] Lu, X., Jin, C., Yang, J., liu, Q., Wu., S., Li., D., Guan., Y., Cai, Y. (2013). Prenatal and lactational lead exposure enhanced oxidative stress and altered apoptosis status in offspring rats' hippocampus. *Biol Trace Elem Res*, 151, 75–84.

[121] Nam, S. M., Ahn, S. C., Go, T. H., Seo, J. S., Nahm, S. S., Chang, B. J., Lee., J. H. (2018). Ascorbic acid ameliorates gestational lead exposure-induced developmental alteration in GAD67 and c-Kit expression in the rat cerebellar cortex. *Biol Trace Elem Res*, 182, 278–286.

[122] Yu, H., Liao, Y., Li, T., Cui, Y., Wang., G., Zhao, F., Jin, Y. (2016). Alterations of synaptic proteins in the hippocampus of mouse offspring induced by developmental lead exposure. *Mol Neurobiol*, 53, 6786–6798.

[123] Gąssowska, M., Baranowska-Bosiacka, I., Moczydlowska, J., Frontczak-Baniewicz, M., Gewartowska, M., Struzynska, L., Gutowska, I., Chlubek, D., Adamczyk, A. (2016). Perinatal exposure to lead (Pb) induces ultrastructural and molecular alterations in synapses of rat offspring. *Toxicology*, 373, 13–29.

[124] Schneider, J. S., Mettil, W. and Anderson, D. W. (2012). Differential effect of postnatal lead exposure on gene expression in the hippocampus and frontal cortex. *J Mol Neurosci*, 47, 76–88.

[125] Luo, M., Xu, Y., Cai, R., Tang, Y., Ge, M. M., Liu, Z. H., Xu, L., Hu, F., Ruan, D. Y., Wang, H. L. (2014). Epigenetic histone modification regulates developmental lead exposure induced hyperactivity in rats. *Toxicol Lett*, 225, 78–85.

[126] Todd, G. D., Wohlers, D. and Citra, M. (2003). Agency for toxic substances and disease registry. Atlanta GA.

[127] IARC Working Group on the Evaluation of Carcinogenic Risks to Humans, World Health Organization, International Agency for Research on Cancer. (2004). Some drinking-water disinfectants and contaminants, including arsenic. IARC.

[128] Smith, A. H., Marshall, G., Yuan, Y., Ferreccio, C., Liaw, J., von Ehrenstein, O., Steinmaus, C., Bates, M.N. and Selvin, S. (2006). Increased mortality from lung cancer and bronchiectasis in young adults after exposure to arsenic in utero and in early childhood. *Environ Health Perspect*, 114, 1293.

[129] Steinmaus, C., Bates, M. N., Yuan, Y., Kalman, D., Atallah, R., Rey, O. A., Biggs, M. L., Hopenhayn, C., Moore, L. E., Hoang, B. K. and Smith, A. H. (2006). Arsenic methylation and bladder cancer risk in case–control studies in Argentina and the United States. *J Occup Environ Med*, 48, 478–488.

[130] Centeno, J. A., Mullick, F. G., Martinez, L., Page, N.P., Gibb, H., Longfellow, D., Thompson, C. and Ladich, E.R. (2002). Pathology related to chronic arsenic exposure. *Environ Health Perspect*, 110, 883.

[131] Karagas, M. R., Punshon, T., Sayarath, V., Jackson, B. P., Folt, C. L., Cottingham, K. L. (2016). Association of rice and rice-product consumption with arsenic exposure early in life. *JAMA Pediatr*, 170, 609–616.

[132] Zhao, F. J., McGrath, S. P. and Meharg, A. A. (2010). Arsenic as a food chain contaminant: Mechanisms of plant uptake and metabolism and mitigation strategies. *Annu Rev Plant Biol*, 61, 535–559.

[133] Zhu, Y., Xi, S., Li, M., Ding, T.T., Liu, N., Cao, F.Y., Zeng, Y., Liu, X.J., Tong, J.W. and Jiang, S.F. (2017). Fluoride and arsenic exposure affects spatial memory and activates the ERK/CREB signaling pathway in offspring rats. *Neurotoxicology*, 59, 56–64.

[134] Zhao, F., Liao, Y., Tang, H., Piao, J., Wang, G., Jin, Y. (2017). Effects of developmental arsenite exposure on hippocampal synapses in mouse offspring. *Metallomics*, 9, 1394–1412.

[135] Xi, S., Sun, W., Wang, F., Jin, Y., Sun, G. (2009). Transplacental and early life exposure to inorganic arsenic affected development and behavior in offspring rats. *Arch Toxicol*, 83, 549–556.

[136] Lupien, S. J. and Lepage, M. (2001). Stress, memory, and the hippocampus: can't live with it, can't live without it. *Behav Brain Res*, 127, 137–158.

[137] Martinez, E. J., Kolb, B. L., Bell, A., Savage, D. D., Allan, A. M. (2008). Moderate perinatal arsenic exposure alters neuroendocrine markers associated with depression and increases depressive-like behaviors in adult mouse offspring. *Neurotoxicology*, 29, 647–655.

[138] Goggin, S. L., Labrecque, M. T. and Allan, A. M. (2012). Perinatal exposure to 50 ppb sodium arsenate induces

hypothalamic-pituitary-adrenal axis dysregulation in male C57BL/6 mice. *Neurotoxicology*, *33*, 1338–1345.

[139] Martinez-Finley, E. J., Ali, A. M. S. and Allan, A. M. (2009). Learning deficits in C57BL/6J mice following perinatal arsenic exposure: consequence of lower corticosterone receptor levels? *Pharmacol Biochem Behav*, 94, 271–277.

[140] Martinez-Finley, E. J., Goggin, S. L., Labrecque, M. T. and Allan, A. M. (2011). Reduced expression of MAPK/ERK genes in perinatal arsenic-exposed offspring induced by glucocorticoid receptor deficits. *Neurotoxicol Teratol*, 33, 530–537.

[141] Pariante, C. M. (2004). Glucocorticoid receptor function *in vitro* in patients with major depression. *Stress*, 7, 209–219.

[142] Aung, K. H., Kurihara, R., Nakashima, S., Maekawa, F., Nohara, K., Kobayashi, T. and Tsukahara, S. (2013). Inhibition of neurite outgrowth and alteration of cytoskeletal gene expression by sodium arsenite. *Neurotoxicology*, 34, 226–235.

[143] Tyler, C. R. and Allan, A. M. (2013). Adult hippocampal neurogenesis and mRNA expression are altered by perinatal arsenic exposure in mice and restored by brief exposure to enrichment. *PloS One*, 8, e73720.

[144] Moore, C. L., Flanigan, T. J., Law, C. D., Loukotková, L., Woodling, K. A., da Costa, G.G., Fitzpatrick, S. C., Ferguson, S.A. (2019). Developmental neurotoxicity of inorganic arsenic exposure in Sprague-Dawley rats. *Neurotoxicol Teratol*, 72, 49–57.

[145] Harada, M. (1995). Minamata disease: Methylmercury poisoning in Japan caused by environmental pollution. *Crit Rev Toxicol*, 25, 1–24.

[146] Atchison, W. D. and Hare, M. F. (1994). Mechanisms of methylmercury-induced neurotoxicity. *FASEB J*, 8, 622–629.

[147] Sarafian, T. and Verity, M. A. (1991). Oxidative mechanisms underlying methyl mercury neurotoxicity. *Int J Dev Neurosci*, 9, 147–153.

[148] Franco, J. L., Teixeira, A., Meotti, F. C., Ribas, C. M., Stringari, J., Pomblum, S. C. G., Moro, Â. M., Bohrer, D., Bairros, A. V., Dafre, A. L. and Santos, A. R. (2006). Cerebellar thiol status and motor deficit after lactational exposure to methylmercury. *Environ Res*, 102, 22–28.

[149] Manfroi, C. B., Schwalm, F. D., Cereser, V., Abreu, F., Oliveira, A., Bizarro, L., Rocha, J. B. T., Frizzo, M. E. S., Souza, D. O. and Farina, M. (2004). Maternal milk as methylmercury source for suckling mice: Neurotoxic effects involved with the cerebellar glutamatergic system. *Toxicol Sci*, 81, 172–178.

[150] Farina, M., Aschner, M. and Rocha, J. B. (2011). Oxidative stress in MeHg-induced neurotoxicity. *Toxicol Appl Pharmacol*, 256, 405–417.

[151] Stringari, J., Nunes, A. K., Franco, J. L., Bohrer, D., Garcia, S. C., Dafre, A. L., Milatovic, D., Souza, D. O., Rocha, J. B., Aschner, M. and Farina, M. (2008). Prenatal methylmercury exposure hampers glutathione antioxidant system ontogenesis and causes long-lasting oxidative stress in the mouse brain. *Toxicol Appl Pharmacol*, 227, 147–154.

[152] Unoki, T., Akiyama, M., Kumagai, Y., Gonçalves, F. M., Farina, M., Da Rocha, J. B. T. and Aschner, M. (2018). Molecular pathways associated with methylmercury-induced Nrf2 modulation. *Front Genet*, 9, 373.

[153] Lu, T. H., Hsieh, S. Y., Yen, C. C., Wu, H. C., Chen, K. L., Hung, D. Z., Chen, C. H., Wu, C. C., Su, Y. C., Chen, Y. W. and Liu, S. H. (2011). Involvement of oxidative stress-mediated ERK1/2 and p38 activation regulated mitochondria-dependent apoptotic signals in methylmercury-induced neuronal cell injury. *Toxicol Lett*, 204, 71–80.

[154] Guida, N., Laudati, G., Mascolo, L., Valsecchi, V., Sirabella, R., Selleri, C., Di Renzo, G., Canzoniero, L. M. and Formisano, L. (2017). p38/Sp1/Sp4/HDAC4/BDNF axis is a novel molecular

pathway of the neurotoxic effect of the methylmercury. *Front Neurosci*, 11, 8.

[155] Kaspar, J. W., Niture, S. K. and Jaiswal, A. K. (2009). Nrf2, INrf2 (Keap1) signaling in oxidative stress. *Free Radic Biol Med*, 47, 1304–1309.

[156] Radonjic, M., Cappaert, N. L., de Vries, E. F., de Esch, C. E., Kuper, F.C., van Waarde, A., Dierckx, R. A., Wadman, W. J., Wolterbeek, A. P., Stierum, R. H. and de Groot, D. M. (2013). Delay and impairment in brain development and function in rat offspring after maternal exposure to methylmercury. *Toxicol Sci*, 133, 112–124.

[157] Markowski, V. P., Flaugher, C. B., Baggs, R. B., Rawleigh, R. C., Cox, C., Weiss, B. (1998). Prenatal and lactational exposure to methylmercury affects select parameters of mouse cerebellar development. *Neurotoxicology*, 19, 879–892.

[158] Jacob, S. and Thangarajan, S. (2018) Fisetin impedes developmental methylmercury neurotoxicity via downregulating apoptotic signalling pathway and upregulating Rho GTPase signalling pathway in hippocampus of F1 generation rats. *Int J Dev Neurosci*, 69, 88–96.

[159] Huang, C. F., Liu, S. H., Hsu, C. J. and Lin-Shiau, S. Y. (2011). Neurotoxicological effects of low-dose methylmercury and mercuric chloride in developing offspring mice. *Toxicol Lett*, 201, 196–204.

[160] Aschner, J. L. and Aschner, M. (2005). Nutritional aspects of manganese homeostasis. *Mol Aspects Med*, 26, 353–362.

[161] Ferraz, H. B., Bertolucci, P. H. F., Pereira, J. S., Lima, J. G. C., Andrade, L. A. F. D. (1988). Chronic exposure to the fungicide maneb may produce symptoms and signs of CNS manganese intoxication. *Neurology*, 38, 550–550.

[162] dos Santos, N. R., Rodrigues, J. L., Bandeira, M. J., Anjos, A. L. D. S., Cecília de Freitas, S. A., Adan, L. F. F. and Menezes-Filho, J. A. (2019). Manganese exposure and association with hormone

imbalance in children living near a ferro-manganese alloy plant. *Environ Res*, 172, 166–174.

[163] da Silva, C. J., da Rocha, A. J., Mendes, M. F., de Mello Braga, A. P. S., Jeronymo, S. (2008). Brain manganese deposition depicted by magnetic resonance imaging in a welder. *Arch Neurol*, 65, 983–983.

[164] Lucchini, R., Apostoli, P., Perrone, C., Placidi, D., Albini, E., Migliorati, P., Mergler, D., Sassine, M.P., Palmi, S. and Alessio, L. (1999). Long-term exposure to "low levels" of manganese oxides and neurofunctional changes in ferroalloy workers. *Neurotoxicology*, 20, 287–297.

[165] Dörner, K., Dziadzka, S., Höhn, A., Sievers, E., Oldigs, H. D. Schulz-Lell, G. and Schaub, J. (1989). Longitudinal manganese and copper balances in young infants and preterm infants fed on breast-milk and adapted cow's milk formulas. *Br J Nutr*, 61, 559–572.

[166] Lucchini, R., Placidi, D., Cagna, G., Fedrighi, C., Oppini, M., Peli, M., Zoni, S. (2017). Manganese and developmental neurotoxicity. In: *Neurotoxicity of Metals*. Springer, pp. 13–34.

[167] Oulhote, Y., Mergler, D., Barbeau, B., Bellinger, D. C., Bouffard, T., Brodeur, M. È., Saint-Amour, D., Legrand, M., Sauvé, S. and Bouchard, M. F. (2014). Neurobehavioral function in school-age children exposed to manganese in drinking water. *Environ Health Perspect,* 122, 1343–1350.

[168] Zoni, S. and Lucchini, R. G. (2013). Manganese exposure: cognitive, motor and behavioral effects on children:a review of recent findings. *Curr Opin Pediatr*, 25, 255.

[169] Bouchard, M., Laforest, F., Vandelac, L., Bellinger, D., Mergler, D. (2006). Hair manganese and hyperactive behaviors: Pilot study of school-age children exposed through tap water. *Environ Health Perspect*, 115, 122–127.

[170] Wasserman, G. A., Liu, X., Parvez, F., Factor-Litvak, P., Ahsan, H., Levy, D., Kline, J., van Geen, A., Mey, J., Slavkovich, V. and

Siddique, A.B. (2011). Arsenic and manganese exposure and children's intellectual function. *Neurotoxicology*, 32, 450–457.

[171] Farina, M., Avila, D. S., Da Rocha, J. B. T., Aschner, M. (2013). Metals, oxidative stress and neurodegeneration: A focus on iron, manganese and mercury. *Neurochem Int*, 62, 575–594.

[172] Erikson, K. M., Dorman, D. C., Fitsanakis, V., Lash, L. H., Aschner, M. (2006). Alterations of oxidative stress biomarkers due to in utero and neonatal exposures of airborne manganese. *Biol Trace Elem Res*, 111, 199–215.

[173] Dobson, A. W., Weber, S., Dorman, D. C., Lash, L. K., Erikson, K. M., Aschner, M. (2003). Oxidative stress is induced in the rat brain following repeated inhalation exposure to manganese sulfate. *Biol Trace Elem Res*, 93, 113–125.

[174] Gunter, T. E., Gerstner, B., Lester, T., Wojtovich, A.P., Malecki, J., Swarts, S.G., Brookes, P.S., Gavin, C.E. and Gunter, K.K. (2010). An analysis of the effects of Mn^{2+} on oxidative phosphorylation in liver, brain, and heart mitochondria using state 3 oxidation rate assays. *Toxicol Appl Pharmacol*, 249, 65–75.

[175] Peres, T. V., Eyng, H., Lopes, S. C., Colle, D., Gonçalves, F. M., Venske, D. K., Lopes, M. W., Ben, J., Bornhorst, J., Schwerdtle, T. and Aschner, M. (2015). Developmental exposure to manganese induces lasting motor and cognitive impairment in rats. *Neurotoxicology*, 50, 28–37.

[176] Amos-Kroohs, R. M., Davenport, L. L., Atanasova, N., Abdulla, Z. I., Skelton, M. R., Vorhees, C. V. and Williams, M. T. (2017). Developmental manganese neurotoxicity in rats: Cognitive deficits in allocentric and egocentric learning and memory. *Neurotoxicol Teratol*, 59, 16–26.

[177] Wang, L., Ohishi, T., Shiraki, A., Morita, R., Akane, H., Ikarashi, Y., Mitsumori, K. and Shibutani, M. (2012). Developmental exposure to manganese chloride induces sustained aberration of

neurogenesis in the hippocampal dentate gyrus of mice. *Toxicol Sci*, 127, 508–521.

[178] Ohishi, T., Wang, L., Akane, H., Shiraki, A., Goto, K., Ikarashi, Y., Suzuki, K., Mitsumori, K. and Shibutani, M. (2012). Reversible aberration of neurogenesis affecting late-stage differentiation in the hippocampal dentate gyrus of rat offspring after maternal exposure to manganese chloride. *Reprod Toxicol*, 34, 408–419.

[179] Cordova, F. M., Aguiar, A. S., Peres, T. V., Lopes, M.W., Gonçalves, F. M., Pedro, D. Z., Lopes, S. C., Pilati, C., Prediger, R. D., Farina, M. and Erikson, K. M. (2013). Manganese-exposed developing rats display motor deficits and striatal oxidative stress that are reversed by Trolox. *Arch Toxicol*, 87, 1231–1244.

[180] Wang, L., Fu, H., Liu, B., Liu, X., Chen, W., Yu, X. (2017). The effect of postnatal manganese exposure on the NMDA receptor signaling pathway in rat hippocampus. *J Biochem Mol Toxicol*, 31, e21969.

[181] Moreno, J. A., Streifel, K. M., Sullivan, K. A., Legare, M. E. and Tjalkens, R. B. (2009). Developmental exposure to manganese increases adult susceptibility to inflammatory activation of glia and neuronal protein nitration. *Toxicol Sci*, 112, 405–415.

[182] Smith, E. A. and Oehme, F. W. (1991). Acrylamide and polyacrylamide:a review of production, use, environmental fate and neurotoxicity. De Gruyter.

[183] Tareke, E., Rydberg, P., Karlsson, P., Eriksson, S., Törnqvist, M. (2002). Analysis of acrylamide, a carcinogen formed in heated foodstuffs. J Agric Food Chem, 50, 4998–5006.

[184] Manson, J., Brabec, M. J., Buelke-Sam, J., Carlson, G. P., Chapin, R. E., Favor, J. B., Fischer, L. J., Hattis, D., Lees, P. S., Perreault-Darney, S. and Rutledge, J. (2005). NTP-CERHR Expert panel report on the reproductive and developmental toxicity of acrylamide. *Reprod Toxicol*, 74, 17–113.

[185] LoPachin, R. M., Ross, J. F., Reid, M. L., Das, S., Mansukhani, S., Lehning, E. J. (2002). Neurological evaluation of toxic axonopathies in rats: Acrylamide and 2, 5-hexanedione. *Neurotoxicology*, 23, 95–110.

[186] Lehning, E. J., Persaud, A., Dyer, K. R., Jortner, B. S., LoPachin, R. M. (1998). Biochemical and morphologic characterization of acrylamide peripheral neuropathy. *Toxicol Appl Pharmacol*, 151, 211–221.

[187] LoPachin, R. M. (2004). The changing view of acrylamide neurotoxicity. *Neurotoxicology*, 25, 617–630.

[188] Wang, H., Huang, P., Lie, T., Li, J., Hutz, R. J., Li, K., Shi, F. (2010). Reproductive toxicity of acrylamide-treated male rats. *Reprod Toxicol*, 29, 225–230.

[189] Rice, J. M. (2005) The carcinogenicity of acrylamide. Mutat Res *Toxicol Environ Mutagen*, 580, 3–20.

[190] Zhang, L., Gavin, T., Barber, D. S., LoPachin, R. M. (2011) Role of the Nrf2-ARE pathway in acrylamide neurotoxicity. *Toxicol Lett*, 205, 1–7.

[191] Calabrese, V., Mancuso, C., Calvani, M., Rizzarelli, E., Butterfield, D. A., Stella, A. M. G. (2007). Nitric oxide in the central nervous system: Neuroprotection versus neurotoxicity. *Nat Rev Neurosci*, 8, 766.

[192] Barber, D. S., Stevens, S. and LoPachin, R. M. (2007). Proteomic analysis of rat striatal synaptosomes during acrylamide intoxication at a low dose rate. *Toxicol Sci*, 100, 156–167.

[193] Duarte-Salles, T., Von Stedingk, H., Granum, B., Gützkow, K. B, Rydberg, P., Törnqvist, M., Mendez, M. A, Brunborg, G., Brantsæter, A. L., Meltzer, H. M., Alexander, J. (2012). Dietary acrylamide intake during pregnancy and fetal growth—results from the Norwegian mother and child cohort study (MoBa). *Environ Health Perspect*, 121, 374–379.

[194] Pedersen, M., Von Stedingk, H., Botsivali, M., Agramunt, S., Alexander, J., Brunborg, G., Chatzi, L., Fleming, S., Fthenou, E.,

Granum, B. and Gutzkow, K. B. (2012). Birth weight, head circumference, and prenatal exposure to acrylamide from maternal diet: The European Prospective Mother–Child Study (New Generis). *Environ Health Perspect*, 120, 1739.

[195] LoPachin, R. M. and Gavin, T. (2012). Molecular mechanism of acrylamide neurotoxicity: Lessons learned from organic chemistry. *Environ Health Perspect*, 120, 1650–1657.

[196] Rodríguez-Ramiro, I., Martín, M. Á., Ramos, S., Bravo, L., Goya, L. (2011). Olive oil hydroxytyrosol reduces toxicity evoked by acrylamide in human Caco-2 cells by preventing oxidative stress. *Toxicology*, 288, 43–48.

[197] Liu, Z., Song, G., Zou, C., Liu, G., Wu, W., Yuan, T., Liu, X. (2015). Acrylamide induces mitochondrial dysfunction and apoptosis in BV-2 microglial cells. *Free Radic Biol Med*, 84, 42–53.

[198] Prasad, S. N. and Muralidhara. (2012). Evidence of acrylamide induced oxidative stress and neurotoxicity in Drosophila melanogaster–Its amelioration with spice active enrichment: Relevance to neuropathy. *Neurotoxicology*, 33, 1254–1264.

[199] Krishna, G. and Muralidhara. (2015). Inulin supplementation during gestation mitigates acrylamide-induced maternal and fetal brain oxidative dysfunctions and neurotoxicity in rats. *Neurotoxicol Teratol*, 49, 49–58.

[200] Krishna, G., Divyashri, G., Prapulla, S. G., Muralidhara. (2015). A combination supplement of fructo-and xylo-oligosaccharides significantly abrogates oxidative impairments and neurotoxicity in maternal/fetal milieu following gestational exposure to acrylamide in rat. *Neurochem Res*, 40, 1904–1918.

[201] Allam, A., El-Ghareeb, A. A., Abdul-Hamid, M., Baikry, A., Sabri, M. I. (2011). Prenatal and perinatal acrylamide disrupts the development of cerebellum in rat: Biochemical and morphological studies. *Toxicol Ind Health*, 27, 291–306.

[202] Erdemli, M. E., Turkoz, Y., Altinoz, E., Elibol, E., Dogan, Z. (2016). Investigation of the effects of acrylamide applied during pregnancy on fetal brain development in rats and protective role of the vitamin E. *Hum Exp Toxicol*, 35, 1337–1344.

[203] Erdemli, M. E., Aladag, M. A., Altinoz, E., Demirtas, S., Turkoz, Y., Yigitcan, B., Bag, H. G. (2018). Acrylamide applied during pregnancy causes the neurotoxic effect by lowering BDNF levels in the fetal brain. *Neurotoxicol Teratol*, 67, 37–43.

[204] Garey, J., Ferguson, S. A. and Paule, M. G. (2005). Developmental and behavioral effects of acrylamide in Fischer 344 rats. *Neurotoxicol Teratol*, 27, 553–563.

[205] Ferguson, S. A., Garey, J., Smith, M. E., Twaddle, N. C., Doerge, D. R., Paule, M. G. (2010). Preweaning behaviors, developmental landmarks, and acrylamide and glycidamide levels after pre-and postnatal acrylamide treatment in rats. *Neurotoxicol Teratol*, 32, 373–382.

[206] Carriel, V., Garzón, I., Campos, A., Cornelissen, M., Alaminos, M. (2017). Differential expression of GAP-43 and neurofilament during peripheral nerve regeneration through bio-artificial conduits. *J Tissue Eng Regen Med*, 11, 553–563.

[207] Ulfig, N., Setzer, M., Neudörfer, F, Bohl, J. (2000). Distribution of SNAP-25 in transient neuronal circuitries of the developing human forebrain. *Neuroreport*, 11, 1259–1263.

[208] Lai, S., Gu, Z., Zhao, M. M., Li, X. X., Ma, Y. X., Luo, L., Liu, J. (2017). Toxic effect of acrylamide on the development of hippocampal neurons of weaning rats. *Neural Regen Res*, 12, 1648.

[209] Seale, S. M., Feng, Q., Agarwal, A. K., El-Alfy, A. T. (2012). Neurobehavioral and transcriptional effects of acrylamide in juvenile rats. *Pharmacol Biochem Behav*, 101, 77–84.

[210] Rosenberg, N. L., Spitz, M. C., Filley, C. M., Davis, K. A., Schaumburg, H. H. (1988). Central nervous system effects of chronic toluene abuse—clinical, brainstem evoked response and

magnetic resonance imaging studies. *Neurotoxicol Teratol*, 10, 489–495.

[211] Boor, J. W. and Hurtig, H. I. (1977). Persistent cerebellar ataxia after exposure to toluene. *Ann Neurol, 2*, 440–442.

[212] Filley, C. M., Halliday, W., Kleinschmidt-DeMasters, B. K. (2004). The effects of toluene on the central nervous system. *J Neuropathol Exp Neurol*, 63, 1–12.

[213] Cruz, S. L., Balster, R. L., Woodward, J. J. (2000). Effects of volatile solvents on recombinant N-methyl-D-aspartate receptors expressed in Xenopus oocytes. *Br J Pharmacol*, 131, 1303–1308.

[214] Bale, A. S., Jackson, M. D., Krantz, Q. T., Benignus, V.A., Bushnell, P.J., Shafer, T.J. and Boyes, W. K. (2007). Evaluating the NMDA-glutamate receptor as a site of action for toluene, in vivo. *Toxicol Sci*, 98, 159–166.

[215] Bale, A. S., Smothers, C. T., Woodward, J. J. (2002). Inhibition of neuronal nicotinic acetylcholine receptors by the abused solvent, toluene. *Br J Pharmacol*, 137, 375–383.

[216] Win-Shwe, T. T., Mitsushima, D., Nakajima, D., Ahmed, S., Yamamoto, S., Tsukahara, S., Kakeyama, M., Goto, S. and Fujimaki, H. (2007). Toluene induces rapid and reversible rise of hippocampal glutamate and taurine neurotransmitter levels in mice. *Toxicol Lett*, 168, 75–82.

[217] Bushnell, P. J., Crofton, K. M. (1999). Neurobehavioral toxicology of organic solvents. *Introd Neurobehav Toxicol Food Environ*, 395–428.

[218] Win-Shwe, T., Yoshida, Y., Kunugita, N., Tsukahara, S., Fujimaki, H. (2010). Does early life toluene exposure alter the expression of NMDA receptor subunits and signal transduction pathway in infant mouse hippocampus? *Neurotoxicology*, 31, 647–653.

[219] Seo, H. S., Yang, M., Song, M. S., Kim, J. S., Kim, S. H., Kim, J. C., Kim, H., Shin, T., Wang, H. and Moon, C. (2010). Toluene

inhibits hippocampal neurogenesis in adult mice. *Pharmacol Biochem Behav*, 94, 588–594.

[220] Hester, S. D., Johnstone, A. F., Boyes, W. K., Bushnell, P. J., Shafer, T. J. (2011). Acute toluene exposure alters expression of genes in the central nervous system associated with synaptic structure and function. *Neurotoxicol Teratol*, 33, 521–529.

[221] Royland, J. E., Kodavanti, P. R. S., Schmid, J. E., MacPhail, R. C. (2011). Toluene effects on gene expression in the hippocampus of young adult, middle-age, and senescent Brown Norway Rats. *Toxicol Sci*, 126, 193–212.

[222] MacPhail, R. C., Farmer, J. D., Jarema, K. A. (2012). Toluene effects on the motor activity of adolescent, young-adult, middle-age and senescent male Brown Norway rats. *Neurotoxicology*, 33, 111–118.

In: A Closer Look at Neurotoxicity
Editor: Gokul Krishna
ISBN: 978-1-53616-591-3
© 2020 Nova Science Publishers, Inc.

Chapter 2

EXTRACELLULAR PROPAGATION OF LIPID RAFT MOLECULAR SPECIES INVOLVED IN PARKINSON'S DISEASE NEURODEGENERATION

Daniel Pereda[1], PhD, Ricardo Puertas-Avendaño[1], PhD, Milorad Zjalic[2], Marija Heffer[2], MD, Ibrahim Gonzalez-Marrero[1], PhD, Miriam González-Gómez[1], PhD, Mario Diaz[3], PhD and Raquel Marin[1,], PhD*

[1]Department of Medical Basic Sciences, School of Medicine, University of La Laguna, La Laguna, Tenerife, Spain
[2]Department of Medical Biology and Genetics, Faculty of Medicine Osijek, Josip Juraj Strossmayer, University of Osijek, Cara Hadrijana, Osijek, Croatia
[3]Laboratory of Membrane Physiology and Biophysics, Department of Animal Biology, University of La Laguna, La Laguna, Tenerife, Spain; Associated Unit IPNA-CSIC-ULL "Physiology and

---

[*]Corresponding Author's E-mail: rmarin@ull.es.

Biophysics of cell membrane in neurodegenerative and tumoral pathologies", Tenerife, Spain

ABSTRACT

Exosomes are small membrane nanovesicles, generally 50 to 90 nm size, secreted by cells upon fusion with the cell membrane. When released, exosomes can propagate molecular content to other cells over long distances. In neurons, these nanovesicles contribute to regulating neuronal development, plasticity and regeneration. Related to neurodegenerative diseases, such as Alzheimer's disease (AD) and Parkinson's disease (PD), exosomes importantly contribute to propagating neurotoxic aberrant protein aggregates and misfolded markers, thus contributing to increase the neurotoxicity impact.

Recent evidence has demonstrated that some of the typical hallmarks of AD and PD, i.e., amyloid- beta peptide (Aβ) and α-synuclein (α-Syn) associate with lipid markers integrated in lipid raft membrane microdomains as part of their mechanisms of pathological self-aggregation. Lipid rafts are laterally organized structures within cell membranes based on a particular lipid composition enriched in gangliosides, sphingolipids and cholesterol. Our previous work has extensively demonstrated that these membrane microstructures are involved in AD, PD and other synucleopathies. Indeed, different lipid species such as ganglioside GM1 and cholesterol highly enriched in lipid rafts, known to enhance Aβ and α-Syn oligomerization are also involved in the propagation of toxic aggregates.

We demonstrate here that Aβ and α-Syn markers are abundantly secreted specifically in 30-90 nm size vesicles. Other protein raft markers known to participate in AD pathology, such as flotillin-1 and the voltage dependent anion channel (VDAC) are also found in these exosomes. Noticeably, this 30-90 nm subclass shows a high content of ganglioside GM1, as an indicative of the potential involvement of this lipid class in toxic protein markers propagation.

Overall, these results suggest that lipid rafts are involved not only in the mechanisms of conformational transition and oligomerization of toxic protein markers but also in the cell to cell propagation of aberrant molecular species that may enhance the neuropathological progression.

Keywords: nanovesicles, exosomes, lipid rafts, alpha-synuclein, amyloid beta peptide, voltage-dependent anion channel 1, Alzheimer's disease, Parkinson's disease

ABBREVIATIONS

ABCA5	ATP-Binding cassette subfamily A transporter 5
APP	Amyloid precursor protein
Aβ	Amyloid beta peptide
ChTx	Cholera toxin
DMEM	Dulbecco's modified eagle medium
FBS	Foetal bovine serum
MPTP	1-methyl-4-phenyl-1,2,3,6-tetrahydropyridine
PD	Parkinson's disease
RA	Retinoic acid
TPA	Phorbol-12-myristate-13-acetate
VDAC-1	Voltage-dependent anion channel 1

INTRODUCTION

In normal non-pathological circumstances, cells maintain control of the biosynthesis, folding, sorting, trafficking and degradation of its proteome through many different mechanisms, a phenomenon known as proteostasis. Whenever this proteostasis is disturbed in the cell, a pathological scenario may appear. Neurodegenerative diseases are a heterogeneous group of chronic disorders that are characterized by the progressive loss of neurons, the gradual degeneration of the structure and the eventual dysfunction of the nervous system. The precise nature of each disorder varies widely, depending on factors such as the brain region and neuronal types affected, the pathogenic protein involved and the time course of cell death. However, a number of common traits appear to be common in most of them. One of such hallmarks is the misfolding and aggregation of proteins (Shastry, 2003; Sweeney et al., 2017), which often results in cytotoxic structures of diagnostic value, e.g., the Lewy bodies in PD and Lewy body dementias, or the amyloid plaques in AD. In the case of AD, the main neuropathological protein hallmarks Aβ and Tau protein adopt pathological configurations which accumulate and spread in the frontal cortex and hippocampus contributing to cognitive decline and dementia (Selkoe 2001). In PD and other synucleopathies, α-Syn protein misfolds and forms toxic aggregates preferentially in dopaminergic neurons leading to motor dysfunction and the typical pathological features of this disease (Beitz 2014).

Since Prusiner's seminal paper on "infectious proteins" (Prusiner, 1982), many proteins have been found that possess more than one biologically active conformation of which at least one is able to self-propagate and form highly stable, non-functional aggregates. In this sense, Aβ is the major component of the amyloid plaques in AD (Murphy and LeVine, 2010). The aggregated fibrils of α-Syn form insoluble Lewy bodies which characterize a superfamily of neurodegenerative diseases known as synucleinopathies. The most common synucleinopathies are PD, dementia with Lewy bodies (DLB) and multiple system atrophy (Barker and Williams-Gray, 2016). Although it has been known for some time that both aggregation-prone proteins may act synergistically in different ways (Marsh and Blurton-Jones, 2012), more recent evidence has shown that they also associate with lipid markers integrated in specific cell membrane domains as part of their mechanisms of pathological self-aggregation (Lemkul and Bevan, 2013; van Maarschalkerweerd et al., 2015).

The most characteristic membrane lipid microstructures are lipid rafts. Lipid rafts are dynamic plasma membrane microdomains enriched in cholesterol, gangliosides and sphingolipids (Brown and London,

2000; Lingwood and Simons, 2010). This peculiar lipid composition allows the clustering of numerous proteins that are arranged in signalling platforms, or signalosomes. Signalosomes are dynamic lipid-protein structures that bind to a variety of extracellular ligands triggering different cellular responses (Levental and Veatch, 2016). Indeed, some data have revealed the importance of neuronal signalosomes in neuroprotective mechanisms of action against AD (Marin, 2011).

The maintenance of lipid raft multimolecular complexes and specific molecular layout in the membrane is possible thanks to a highly ordered, high viscosity microstructure that facilitates the anchoring of complex molecules, such as glycosylphosphatidylinositol (GPI), and protein chemical modifications, such as myristoylation and palmitoylation (Fantini, 2007; Levental et al., 2010).

More importantly for the purpose of this chapter, the chemical composition of lipid rafts shows striking similarities (an abundance of cholesterol and sphingolipids, GPI-anchored proteins and the raft marker flotillin) to that of exosomes. Exosomes are membrane nanovesicles, typically 30 to 90 nm in diameter, secreted by cells upon the fusion of a multivesicular endosome with the plasma membrane (Raposo and Stoorvogel, 2013). Exosomes are a powerful tool for interneuronal communication (Chivet et al., 2012) that can even involve the endosomal pathway of recipient neurons to spread their cargo through the central nervous system (CNS). These vesicles may content a variety of molecular material, including self-propagating and pathological proteins (Polanco et al., 2018). For instance, both the amyloid precursor protein (APP), whose proteolysis generates Aβ, and α-Syn have been found in exosomal fractions of the central nervous system (Danzer et al., 2012; Xiao et al., 2017).

Apart from cholesterol and sphingolipids, another relevant lipid component of lipid rafts is the ganglioside GM1, a multitasking molecule involved in many signalling pathways that has been claimed as a promising therapeutic tool for the treatment of synucleinopathies

(Schneider et al., 2010; Schneider et al., 2015). GM1 is highly abundant in lipid rafts and thus is a very useful lipid raft marker for the study of these microdomains (Ledeen and Wu, 2015). Furthermore, GM1 is a main target of self-aggregating molecules such as α-Syn and Aβ. Indeed, different results have demonstrated a role of GM1 in the control of α-Syn trafficking into the plasma membrane that may determine the functionality of this protein in the brain (Fink, 2006; Martinez et al., 2007; Park et al., 2009). Moreover, GM1 also interacts with Aβ resulting in the aggregation of amyloid fibrils at the vicinity of neuronal lipid rafts, a phenomenon that may regulate the onset of neurodegenerative diseases (Yamamoto et al., 2012).

Our previous work has shown that lipid raft microstructures exhibit pathological modifications leading to the progression of AD, PD and other synucleinopathies (Marin et al., 2016; Canerina-Amaro et al., 2017; Diaz et al., 2018; Marin and Diaz, 2018). In particular, lipid rafts exhibit altered raft lipidomic profiles, thereby modifying their physico-chemical properties that affect protein associations (Marin et al., 2017). Furthermore, some preliminary observations have shown that some raft-associated proteins may be involved in the release and trafficking of exosomes (Phuyal et al., 2014), and in the cell-cell propagation of toxic Aβ and α-Syn aggregates (Badawi et al., 2018).

In the present study, we have explored the relationship between exosomes and lipid rafts in the context of neurodegenerative diseases using the human derived SH-SY5Y cell line, a well-known model of neuronal function, and the neurotoxic drug 1-methyl-4-phenyl-1,2,3,6-tetrahydropyridine (MPTP). MPTP has a colorful and distinguished history as an inductor of reproducible PD-like syndrome across the mammalian lineage (Meredith and Rademacher, 2011). In its native form, the drug crosses the blood brain barrier and it is metabolized into MPP+ in astrocyte lysosomes, which is in turn avidly accumulated by the dopamine transporter in dopaminergic neurons of the *Substantia Nigra pars compacta* (Marini et al., 1992). MPP+ is a pleiotropic neurotoxin that has been characterized by a plethora of distinct intracellular effects that affect, among others, mitochondrial functioning. For instance, this neurotoxin: 1) causes the depletion of ATP by inhibiting the mitochondrial complex I (Nicklas et al., 1985; Chan et al., 1991); and 2) induces the generation of reactive oxygen species (ROS) (Zang and Misra, 1993). Furthermore, an increasing number of reports has demonstrated some harmful effects of MPP+ related to inflammation (Meredith et al., 2005) and excitotoxicity (Meredith et al., 2009). Altogether, these findings reflect that MPTP induces specific and dose-dependent death of dopaminergic neurons in the *Substantia Nigra* and thus it represents a gold standard model for PD *in vivo* and *in vitro*.

We have therefore used this model to show evidence that lipid raft integrated components are part of the propagation mechanisms of Aβ and of α-Syn spreading. The pathophysiological relevance of these phenomena to AD and PD pathologies is discussed herein.

MATERIAL AND METHODS

Materials

MPTP-HCl and cholera toxin B subunit-HRP were purchased from BioSigma (Tenerife, Spain). The mouse polyclonal antibody against aggregated α-Syn was purchased from Millipore (Madrid, Spain). The mouse monoclonal antibody against α-Syn, mouse monoclonal anti-VDAC-1 antibody and the rabbit polyclonal antibody against flotillin-1 were from Abcam (Cambridge, UK). Mouse monoclonal anti-Aβ, goat polyclonal, rat monoclonal anti-DAT antibodies and mouse polyclonal antibodies were purchased from Santa Cruz biotechnologies (Texas, EE.UU). Rabbit anti-ABCA5 and TAPA1 (anti-CD81) and mouse anti-CD63 and anti-ICAM 1 antibodies were purchased from Abcam (Cambridge, United Kingdom). PreCast Mini-Protean SDS-PAGE gels,

Transblot® Turbo™ Mini PVDF Transfer Packs, and Clarity™ Western ECL substrate were from Bio-Rad Laboratories (Madrid, Spain). The exosome extraction kit was from Thermo Fisher Scientific (Madrid, Spain). The inclusion Spur resin for electron microscopy was purchased from Aname (Madrid, Spain).

Cell Culture and Treatments

Human neuroblastoma SH-SY5Y cells (American Type Culture Collection; ATCC, Rockville, MD) were grown in Dulbecco's modified Eagle's medium/F-12 medium, containing 10% fetal bovine serum exosome-free Somelglas (Madrid, Spain), 0.58 g/l glutamine, 3.7 g/l $NaHCO_3$, and 1% penicillin/streptomycin. Cells were cultured at 37°C under 5% CO_2-95% atmosphere.

To generate fully differentiated dopaminergic phenotype, SH-SY5Y cells were differentiated with all-trans-retinoic acid and phorb-ol ester 12-O-tetradecanoylphorbol-13-acetate (RA/TPA), as previously described (Presgraves et al., 2004) with minor modifications. Briefly, cells at 40% confluency were exposed to retinoic acid (10 mM) for 5 days. Then the media were removed and replaced with fresh TPA medium (320 nM) for another 2 days.

For MPTP treatment (a dopaminergic neurotoxin), cells were incubated with freshly prepared sterilized stocks (500 mM), and diluted (1:500) into SH-SY5Y cultured medium. Cells were cultured in the presence of 0.75 mM of MPTP for 24 hours, and cell viability was assessed by trypan blue exclusion, as described below. Differentiated SH-SY5Y cells without MPTP treatment were used as controls. All experiments were performed at least three times.

Cell viability was assessed by the trypan blue dye exclusion method (Black and Berenbaum 1964) followed by cell counting. Viable cells (trypan blue-excluding) were counted on a Neubauer haemocytometer.

Data were expressed as the mean percentage of total viable cells relative to the untreated control cultures.

Microscopic Examination

All cultures were examined by phase contrast microscopy (Leica MC170 HD, Berlin) to visualize cell morphology. Representative images were captured using a digital camera (Leica).

Exosomal Fraction Preparation

For exosomal extraction, cells were previously cultured in exofree FBS medium (Solmeglas, Madrid, Spain), and differentiated for six days with RA/TPA. Culture medium was removed, and the cells were resuspended using a cell scraper. After the removal of cell debris by low-speed centrifugation, exosome isolation was performed either 1) using total exosome isolation reagent from cell culture media (Thermofisher Scientific, Massachussets, USA) or 2) via a serial ultracentrifugation protocol.

Briefly, for the first method, the volume of the resultant cell-free culture media was mixed with 0.5 volumes of the kit solution and incubated overnight at 5°C. After incubation, the samples were centrifuged at $10,000g$ for 1 hour at 5°C and the supernatant discarded. The exosome-containing pellet was resuspended in 10 volumes of 1x PBS buffer and preserved at 5°C for analysis.

For the second method, the required volume was centrifuged at $1200g$ for 10 minutes and then at $10,000g$ for one hour. The pellet was resuspended in 400 μL PBS and ultracentrifuged at $110,000g$ for 2-3h. The resulting, exosome-containing pellet (P4) was again resuspended in 400 μL and stored at -80°C for analysis.

Immunoblotting

Proteins resolved by SDS-PAGE were transferred to polyvinyl (PVDF) membranes using the trans-Blot Turbo rapid western blotting transfer system (Bio-Rad, Madrid, Spain). PVDF membranes were incubated overnight at 4ºC with the different primary antibodies (raft marker flotillin-1, anti-α-Syn, anti-Aβ, APP and anti-VDAC-1.) All antibodies were diluted 1:1,000 in BLOTTO, except antibodies purchased from Santa Cruz biotech that were diluted 1:200 according to manufacturer's indications. Membranes were washed three times for 5 minutes in Tris-buffered saline (TBS) with 0.5% Tween-20. Proteins were detected using the corresponding peroxidase-conjugated anti-mouse or anti-rabbit secondary antibody diluted 1:5,000 for 1h at room temperature. Specific bands were developed with Clarity™ Western ECL Substrate and processed using Chemie-Doc MP Imaging System (Bio-Rad), and analyzed using the Image Lab software.

Immunocytochemistry and Confocal Microscopy

To immunolocalize the different protein markers of the study, cultured SH-SY5Y grown in 8-well cultured slides (Falcon, Madrid, Spain) were fixed using 2% ρ-formaldehyde, 0.5% glutaradehyde and 120 mM sucrose in phosphate-buffered saline (PBS, pH 7.4 with 0,.01% Nonidet P-40 detergent). For lipid raft visualization, cells were chilled on ice for 10 min and washed twice with ice-cold PBS. Then, cells were incubated with 2 µg/ml cholera-toxin B conjugated to Alexa Fluor 555 (Molecular Probes, Madrid, Spain) in 0.1% BSA-PBS for 20 minutes at 4°C, followed by fixation. For double immunofluorescence staining, after fixation, cells were incubated overnight at 4°C with a combination of two compatible monoclonal and polyclonal primary antibodies directed against the different protein markers (diluted 1:500 in PBS) in the presence of 1:200 normal serum. The day after, cells were washed with PBS, and incubated with the secondary biotinylated anti-rabbit antibody and with cyanin-3 coupled anti-mouse antibody, both diluted 1:200 in PBS for 2 h at room temperature. Slides were washed three times in PBS and exposed to cyanine-2-conjugated streptavidin (diluted 1:500 in PBS) for 30 minutes at room temperature. After washing in PBS, slides were cover-slipped and processed for fluorescent detection in a laser scanning confocal imaging Fluoview 1000 system (Olympus, Barcelona, Spain).

Electron Microscopy

Secreted membrane nanovesicles were isolated using the exosomal extraction kit (ThermoFisher Scientific). Electron microscopy was performed as described previously (Théry et al., 2006) with some minor modifications. Briefly, vesicles were fixed with 2% ρ-formaldehyde diluted in 100 nM Na_2HPO_4 for 20 min. Exosomal samples were placed onto formvar-carbon copper coated grids. For colloidal gold double staining, grids were incubated for 1 h at room temperature with secondary antibodies with 6, 10 and 15 nm-size gold beads. After incubation, grids were negatively stained with 2% uranyl acetate. Microphotographs were obtained using a Zeiss EM910 electron microscope at 80 kV (Fa, Prscan Elecktronische Systeme GmbH, Lagerlechfeld, Germany). Digital images were obtained with a Slow Scan CCD-Camera for TEM, 1024x1024 (Olympus Soft Imaging Solutions GmbH, Munster, Germany).

Data Analysis

Quantitative results are expressed as mean ± SEM. Data were analysed by statistical software GraphPad Prism 6 for Windows as deemed appropriate for every experiment.

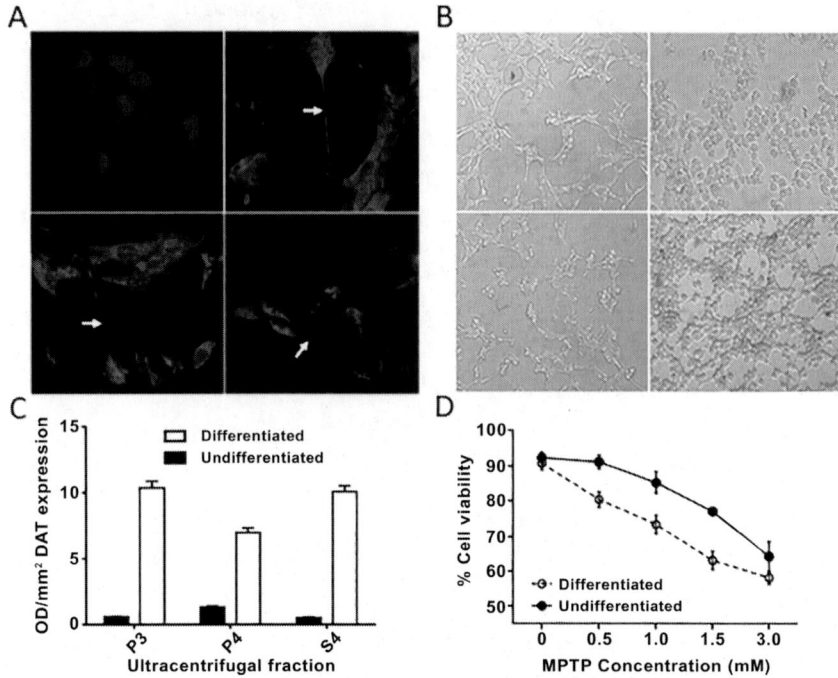

Figure 1. Effect of MPTP in undifferentiated and differentiated SHSY-5Y culture cells. MPTP exposure causes cytotoxicity in SHSY-5Y human neuroblastoma cultured cells depending on whether they were undifferentiated or differentiated cells. SHSY-5Y cells at 40% confluence were incubated into differentiation with RA and TPA and then treated with 0.75 mM MPTP (see MATERIAL AND METHODS for further details). **A**: Immunocytochemical analysis of the expression of DAT (red) in undifferentiated (left) and differentiated (right) SHSY-5Y cells. Nuclei stained with DAPI (blue). **B**: Optical microscopy images of differentiated (top) and undifferentiated (bottom) SHSY-5Y cells before (left) and after (right) treatment with 0.75 mM MPTP. **C**: Optical density analysis of the expression of DAT in segregated ultracentrifugation fractions of culture media of undifferentiated and differentiated SHSY-5Y cells. Data are presented as the mean values ± SEM (n = 6) and statistically significant differences between groups are marked ($^*p < 0.05$, t-test comparison corrected for multiple comparisons using the Holm-Sidak method). **D**: Cell viability assay of undifferentiated and differentiated SHSY-5Y treated with increasing concentrations of MPTP. Data are presented as the mean values ± SEM (n = 4) and statistically significant differences between groups are marked ($^*p < 0.05$, t-test comparison corrected for multiple comparisons using the Holm-Sidak method).

RESULTS

SH-SY5Y Differentiation Induces Dopaminergic Phenotype and Neurotoxicity Vulnerability

Differentiated SH-SY5Y human neuroblastoma cells have long been used as a standard model for the study of neuroprotection and neurotoxicity (Farooqui, 1994; Presgraves et al., 2004). In order to validate whether cell differentiation treatment may modify SH-SY5Y dopaminergic phenotype, culture cells were differentiated with RA/TPA as described in MATERIAL AND METHODS. The differentiation treatment significantly increased dopamine transporter (DAT) immunoreactivity (Figure 1A, top; Figure 1C) as well as the cellular phenotype, observing both a marked α-Syn and flotillin response (Figure 1A, bottom) and a higher neurite density in RA/TPA treated cells as compared to controls (Figure 1A). Moreover, differentiated SHSY-5Y cells showed a higher sensitivity to the MPTP-induced neurotoxicity as compared to undifferentiated controls (Figure 1B, 1D). In view of these results, we performed the rest of experiments on lipid raft markers and exosomal isolates in differentiated cells.

MPTP Treatment Affects the Organization of Lipid Raft and Protein Markers Related to PD-like Pathology in the Plasma Membrane

Many signaling molecules and markers of anatomopathology are known to concentrate in lipid rafts. These multimolecular associations are disrupted by the induction of neurotoxic insults.

To assess whether MPTP neurotoxicity may affect lipid raft markers and the production and conformation of Aβ and α-Syn, we next performed immunocytochemical staining of the lipid raft marker flotillin, α-Syn and Aβ peptide, APP and the voltage-dependent anion channel (VDAC-1), whose dysfunction has been linked to PD and AD (Chu et al., 2014; Smilansky et al., 2015).

Undifferentiated cells showed a mild immunoreactivity for our marker with slight changes in response to MPTP treatment (Figure 2). However, RA/TPA SH-SY5Y cells showed a significant increase in immunoreactivity for α-Syn, flotillin and VDAC-1 as compared to undifferentiated controls (Figure 3). Interestingly, α-Syn and VDAC-1 immunoreactivity were drastically decreased following MPTP treatment (Figure 3). These results suggest that MPTP neurotoxicity induces change in the dynamic of these protein markers at the plasma membrane.

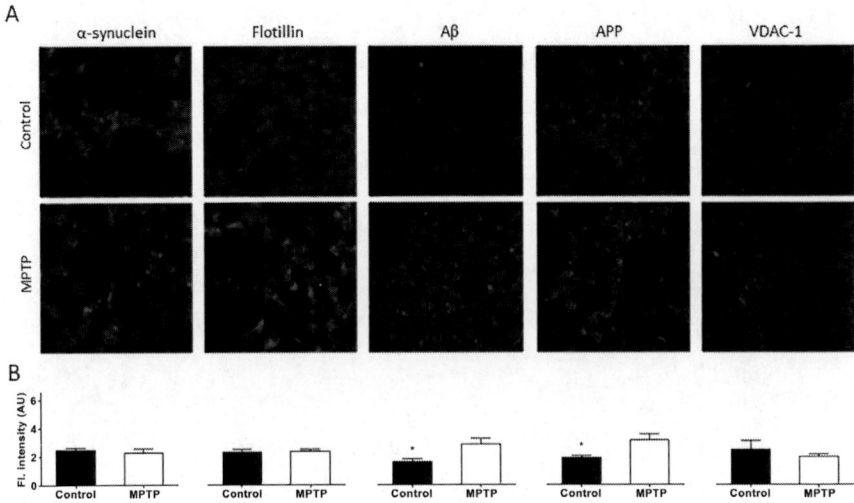

Figure 2. Effect of MPTP treatment in the expression of lipid-raft and protein markers related to pathologies in undifferentiated SHSY-5Y cells. MPTP causes specific changes in the expression of several protein markers in undifferentiated SHSY-5Y cells. **A:** Immunocytochemical staining of undifferentiated cells with MJFR1, flotillin-1, the β-amyloid peptide, APP and VDAC1. A set of cultures were treated with 0.75 mM MPTP neurotoxin and compared with untreated cells. Proteins are stained in red fluorescence and the cellular nuclei stained with DAPI in blue. **B:** Fluorescence intensity quantification of the expression of each protein. ($^*p < 0.05$, t -test comparison with Welch's correction).

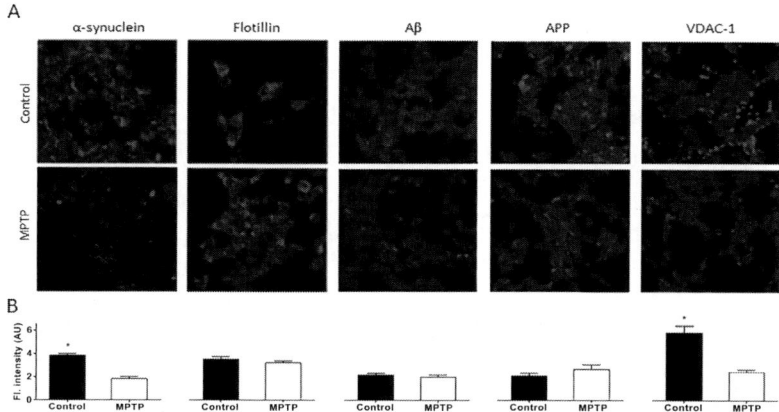

Figure 3. Effect of MPTP treatment in the expression of lipid-raft and protein markers related to pathologies in differentiated SHSY-5Y cells. MPTP causes specific changes in the expression of several protein markers in RA/TPA differentiated SHSY-5Y cells. **A:** Immunocytochemical staining of differentiated cells with MJFR1, flotillin-1, Aβ, APP and VDAC1. A set of cultures were treated with 0.75 mM MPTP neurotoxin and compared with untreated cells. Proteins were stained in red fluorescence and the cellular nuclei stained with DAPI in blue. **B:** Fluorescence intensity quantification of the expression of each protein. ($^*p < 0.05$, t-test comparison with Welch's correction).

MPTP Treatment Causes Extracellular Exosome Secretion

We further investigated whether MPTP treatment may induce lipid raft rearrangements in SHSY-5Y cells. Confocal images of cholera toxin (ChTx)-fluorescent probe as a marker of lipid rafts showed the presence of a punctuated pattern at the neuronal membrane in both, untreated and MPTP-treated cultures (Figure 4A). Furthermore, the analyses by confocal microscopy revealed an increased number of microvesicles ranging from 100 nm to 1 μm in diameter secreting into the extracellular medium as a result of MPTP exposure (Figure 4A).

We next analyzed whether the MPTP-induced toxicity was accompanied by alterations in extracellular secretion. We first characterize the purity of the exosomal extractions segregated by ultracentrifugation of SHSY-5Y culture media, as described in MATERIALS AND METHODS. The exosomal extraction rendered two fraction pellets (P3 and P4) that were analyzed by immunoblotting assays, using specific antibodies against the CD81 exosomal marker (Figure 4B, top). The results showed that the fraction pellet 4 (P4) was highly enriched in the CD81 exosomal marker, whereas this marker was virtually absent in the other isolates (Figure 4B, bottom). It is worth mentioning that fraction pellet 3 (P3) also showed microvesicles content, but they were larger than 100 nm and CD81-negative (data not shown).

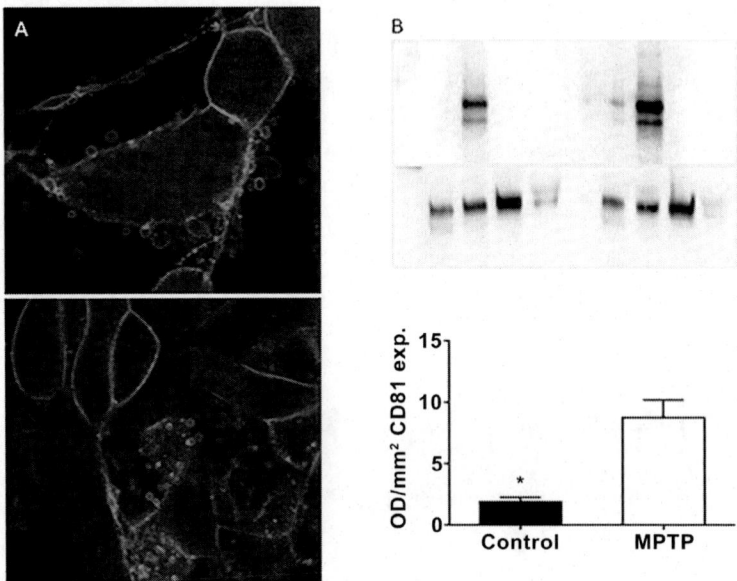

Figure 4. Characterization of lipid rafts and exosomal fractions in differentiated SHSY-5Y cells treated with MPTP. **A**. Lipid rafts stained with Cholera toxin B in green and cell nuclei stained with DAPI in blue. Both control (top) and MPTP-treated (bottom) cells show abundant microvesicles of different sizes (100 nm – 1μm) secreting into the extracellular medium (white arrows). White bar = 25 μm. **B**. (Top) Immunoblotting of P3 (pellet 3), P4 (pellet 4), S4 (supernatant 4) and me (microsomal extracts) obtained by ultracentrifugation of culture media of control (left) and MPTP-tread (right) cells. P4 fractions were enriched in CD81 exosomal marker. (Bottom) Densitometric quantification of CD81 immunoblotting signals indicated a higher abundance of this marker in MPTP-treated cells as compared to control ($*p < 0.05$, Mann-Whitney test).

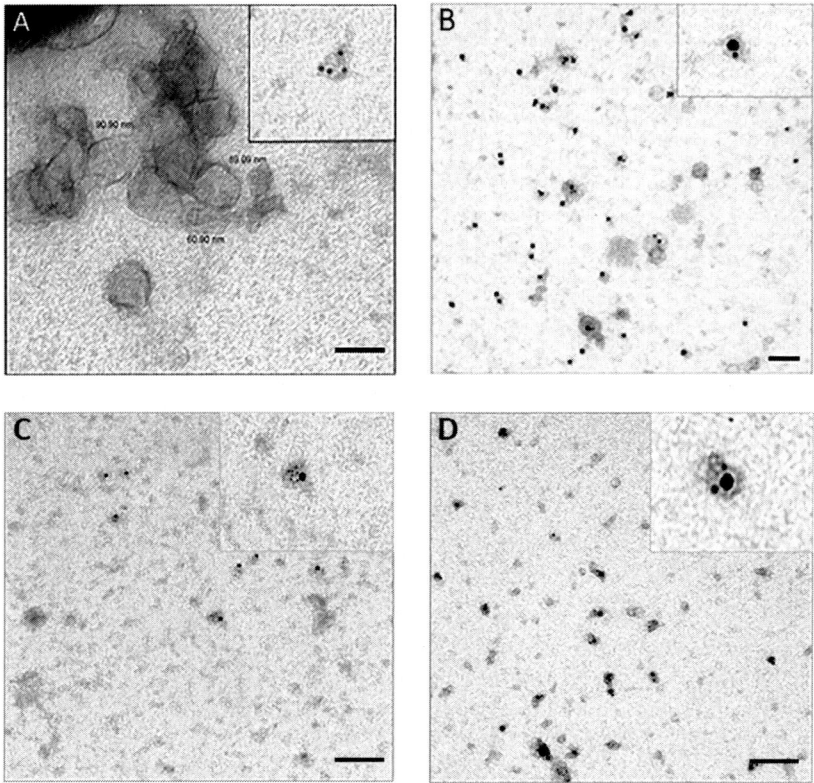

Figure 5. Electron microscopy images of exosomes from the p4 extracts of differentiated MPTP-treated cells **A**. CD81 immunogold staining (10 nm) (inset) showing the typical cup-like shape. Black bar = 100 nm. Exosomes in the range of 20 nm can be detected (insets). **B**. Double-immunogold staining with ICAM1 (15 nm) and Aβ (6 nm). Black bar = 100 nm. **C**. Double-immunogold staining with CD63 (15 nm) and α-Syn (6 nm). **D**. Double-immunogold staining with CD81 (6 nm) and VDAC1 (15 nm).

To confirm this finding, we examined the presence of CD81 in P4 isolates by transmission electron microscopy. We observed a high number of typical cup-shaped exosomes with a diameter between 50 to 100 nm that we could identify with CD81 and ICAM1 immunogold staining (Figure 5A and 5B). ICAM1 is another well characterized exosome marker that was shown to colocalize with α-Syn and VDAC1 by double-immunostaining studies (Figure 5C and D). These results suggest that protein markers involved in neurodegeneration can self-propagate and spread in exosomes.

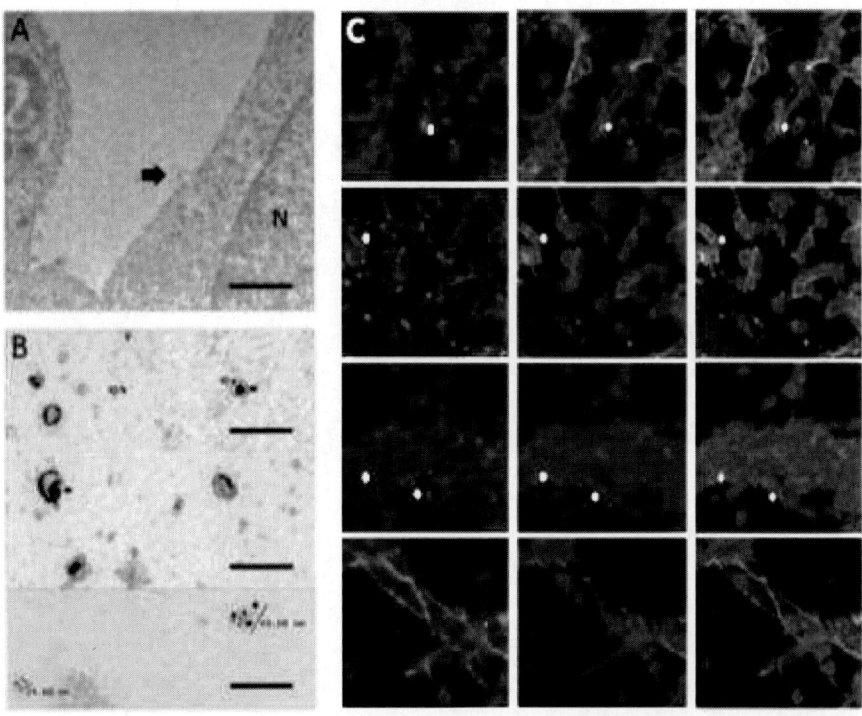

Figure 6. Ganglioside GM1 colocalization with protein markers related to pathology in exosomes. **A**: Extracellular microvesicle (EV, black arrows) secretion in SHSY-5Y cells, following MPTP-treatment. N = cell nucleus. Black bar = 500 nm. **B**: (Top) Double-immunogold staining of exosomes in the fraction P4 of MPTP-treated SHSY-5Y cells with CD63 (15 nm, solid arrows) and flotillin (6 nm, hollow arrows), as well as larger, cup-shaped nanovesicles. Both markers could be seen colocalizing in the same exosomes (inset). Black bar = 100 nm. (Bottom) Double-immunogold staining of exosomes in fraction P4 of MPTP- treated SHSY-5Y cells with GM1 (6 nm) and ABCA5 (10 nm). Arrowheads indicate extracellular vesicles carrying GM1 only (white arrowhead) or co-transporting GM1/ABCA5 (black arrowhead). Black bar = 100 nm. Calculated diameters are included as further reference of size. **C**: Top to bottom row: Flotillin, Aβ, APP and ABCA5 (red), with GM1 (green) and DAPI (blue) staining of untreated SHSY-5Y cells. Merged images show colocalization in yellow (white arrows).

Role of Gangliosides for Extracellular Redistribution of Lipid Raft Protein Markers in the Pathological Scenario

The ganglioside GM1 is abundantly found in lipid rafts where it participates in the clustering and conformation of numerous proteins integrated in these microstructures. For instance, it has been reported that the ganglioside GM1 interacts with α-synuclein and other proteins involved in neurodegeneration (Park et al., 2009; Botto et al., 2014). This interaction regulates the configuration of α-Syn at the neuronal membrane inhibiting its fibrillation (Martinez et al., 2007).

In this order of ideas, it is conceivable that the alteration of lipid rafts observed following MPTP neurotoxicity may affect the extravesicular regulation and lipid raft-related cargo material transported in these vesicles. We therefore performed additional electron microscopy experiments in order to visualize the exosomal secretion following MPTP-treatment. We observed that the vesicle formation was preserved in the presence of the neurotoxin (Figure 6A). Furthermore, double-immunogold staining allowed us to visualize the ganglioside GM1 inside exosomes cotransported with the raft-associated proteins, such as ABCA5 and flotillin-1 (Figure 6B). These results suggest that at least part of the molecular material of lipid raft microstructures is included in exosomal formation.

We further studied by double immunostaining and confocal microscopy the pattern of distribution of GM1 in relation to the proteins of interest in this cell model. The immunocytochemical analysis revealed that the ganglioside GM1 colocalized abundantly with flotillin-1, Aβ, APP and ABCA5 (Figure 6C). Moreover, other ganglioside species known to participate in cellular membrane integrity, such as GD1b and GT1b also showed abundant colocalization with flotillin, Aβ, APP and ABCA5, while none of the gangliosides appear to be associated with α-Syn (see supplementary material). These observations indicate that α-Syn requires the specific presence of GM1, and no other ganglioside species, to be anchored within the plasma membrane.

DISCUSSION AND CONCLUSION

A hallmark of neurodegenerative diseases such as AD and PD is the protein misfolding and propagation of proteins by the prion principle transmission. Here, we observed that a number of proteins involved in the most prevalent neurodegenerative diseases are co-secreted in exosomes, and extracellular nanovesicles released by many cellular types including neurons that can spread their cargo through long distances inside the body (Edgar, 2016). This propagation appears to be promoted by molecular markers of lipid rafts. Furthermore, we have proven that this co-secretion 1) is cell type-dependent and 2) is altered in a statistically significant way in response to PD-like cytotoxicity, as induced by MPTP treatment.

We were particularly interested in the manner that the induced pathological conditions may affect the normal working of the exosomes related to the selected protein markers of this study. In this sense, the high α-Syn immunoreactivity shown by the cells under every experimental condition was as expected, given the relatively high abundancy of this protein in neural cells. Furthermore, the fact that oligomeric α-Syn appeared in the ultracentrifugation fraction S4 of exosomal extraction whereas showing little response to the MPTP treatment, and in correlation with flotillin-1, reinforces the idea that lipid rafts could serve as a scaffold for α-Syn oligomerization. The potential fibril formation by the interaction of this protein with different lipid raft molecular species may reflect the pathological consequences (Zhu and Fink, 2003; Pirc and Ulrih, 2011) and matches the lipid raft fragmentation induced by α-synuclein (Emanuele et al., 2016). The significant detriment of α-synuclein immunoreactivity observed in

MPTP-treated cells points to an effect of the induced PD-like pathology on the structural stability of this protein in the membrane.

An interesting finding was that the cellular differentiation enhanced the segregation of markers associated with plasma membrane turnover. Thus, the ABCA5 transporter involved in the regulation of sphingomyelin levels was shown to be cotransported in extracellular vesicles together with ganglioside GM1. Lastly, a physiological link between α-Syn and the ABCA5 transporter via the cellular sphingomyelin levels has been established (Kim and Halliday, 2012), suggesting an involvement of this protein in α-Syn misfolding in exosomes (Grey et al., 2015). Moreover, our findings correlate with recent biophysical data showing that pathological α-Syn oligomers alter the ability of this protein to interact with lipid membranes dependant upon their chemical composition, to the point that it may no longer be able to bind to microvesicles (Gallea et al., 2018).

These results shed some light on the potential role of exosomes as vehicles for the propagation of pathological protein species. The fact that all the protein markers tested here shown a strong colocalization with GM1 suggests the importance of this ganglioside in extracellular vesicle regulation. The exosomal storage and secretion of proteins involved in neurodegeneration may be, at least, partially a random process in which lipid rafts may be affected by biophysical and biochemical protein and lipid changes. Indeed, the biogenesis and/or the secretion of exosomes themselves might be modified when the rafts microdomains are altered as a consequence of neuronal ageing and enhanced by pathological events (Marin et al., 2017).

If this is the case, it would feasible that a physiological stimulus that disrupts the structural integrity of the lipid rafts could increase the number of exosomes containing lipid raft lipid and protein material in the culture media. MPTP treatment has been shown to cause precisely this disruption in the cell membrane of differentiated SHSY-5Y cells (Eriksson et al., 2017). Indeed, we have detected changes in exosomal content as a consequence of the induced PD-related neurotoxicity, as confirmed by the higher abundance of the CD81 exosomal marker in fraction P4.

According to these observations, we hypothesize the existence of a random sorting of lipid raft lipid and protein material into exosomes facilitated by distortions of the lipid raft microdomains. In accordance to this, some lipid raft scaffolding proteins such as caveolin-1 can induce drastic structural changes in the plasma membrane whose fine tuning has been shown to be affected by a number of different potential alterations in the cellular environment (Marin, 2011).

Though the presence of gangliosides in exosomes, in particular GM1, has been known for some time (Yuyama et al., 2008). For instance, we have observed here that gangliosides can be co-stored in exosomes and co-released with other protein markers related to neuropathology. The aforementioned increase in the number of exosomes due to physiological alterations of the lipid rafts may therefore cause a double detrimental effect in GM1 function: the formation of insoluble toxic aggregates with Aβ that propagate to adjacent cells and the loss of GM1 content in lipid rafts, adding even more structural instability in a potentially pathological descending loop.

We have previously found that brain lipid rafts lipid and protein profiles are altered from the early stages of AD (Fabelo et al., 2014). In particular, early alterations in lipid profiles of these microstructures promote the accumulation of the enzymes related to the processing of Aβ, such as APP and β-secretase (BACE). Interestingly, lipid membrane impairment is also observed during physiological changes occurring with ageing, such as menopause (Canerina-Amaro et al., 2017). Therefore, it is plausible to support a model in which aberrant alterations of the vulnerable lipid raft structure, either induced by pre-existing or novel conditions, may increase the propagation of cytotoxic species related to lipid rafts in other healthy cells thus contributing to the onset of the disease. We have summarized the putative correlation of aberrant lipid raft progression during neurodegeneration that may be reflected in exosomal molecular material (Figure 7). The early events of misfolded aggregates related to lipid rafts may be reflected in the propagation of neuropathological features in exosomes. Feature characterization of this membrane microstructure and exosome cross-link may be applied to future therapies against these diseases.

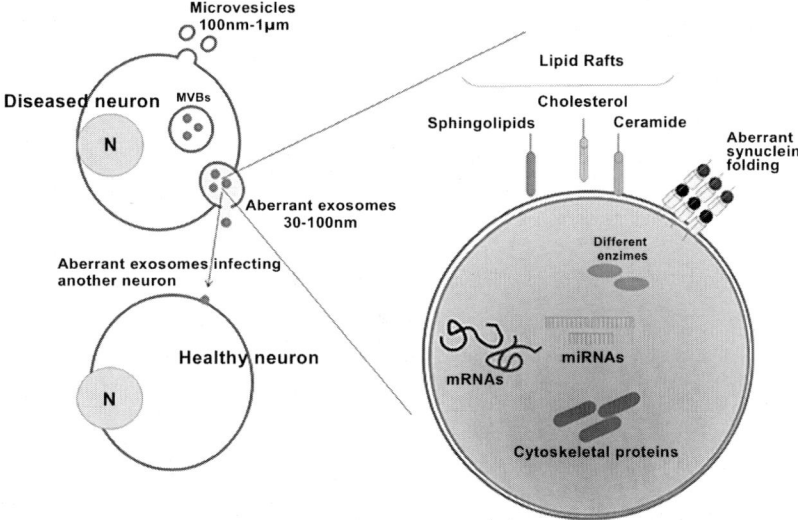

Figure 7. A model of neuropathological markers' propagation related to lipid rafts. Microvesicles of different sizes (30-100 nm) carry genetic material from the cytoplasm and lipid and protein material from lipid rafts in multivesicular bodies to the membrane along other molecules. This material may be released to the extracellular compartment reaching and fusing with other neurons. Aberrant alterations of lipid and protein lipid-raft species resulting from neurotoxicity may increase the propagation of these pathological aggregates inside exosomes thus infecting healthy neurons.

ACKNOWLEDGMENTS

Supported by SAF-2014-52582-R and SAF2017-84454-R (MINECO, Spain).

APPENDIX

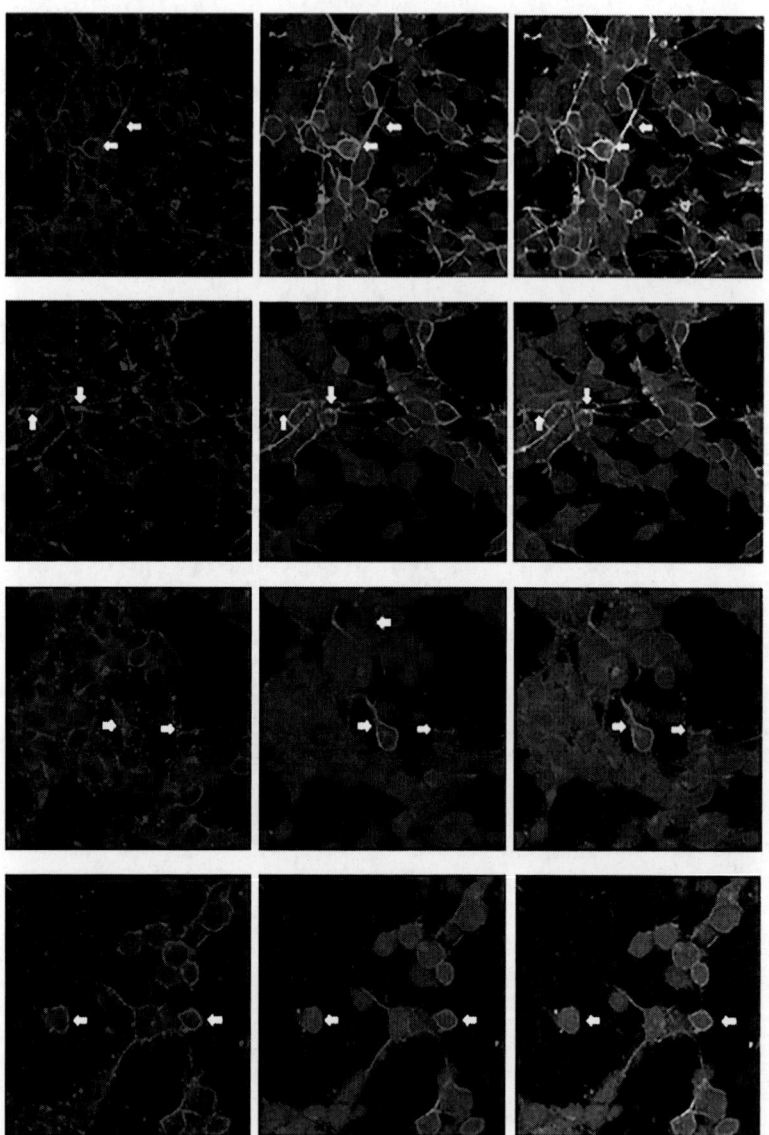

A1. **Ganglioside GD1b colocalization with protein markers related to pathology in exosomes.** Top to bottom row: Flotillin, Aβ, APP and ABCA5 (red), with GD1b (green) and DAPI (blue) staining of untreated SHSY-5Y cells. Merged images show colocalization in yellow (white arrows).

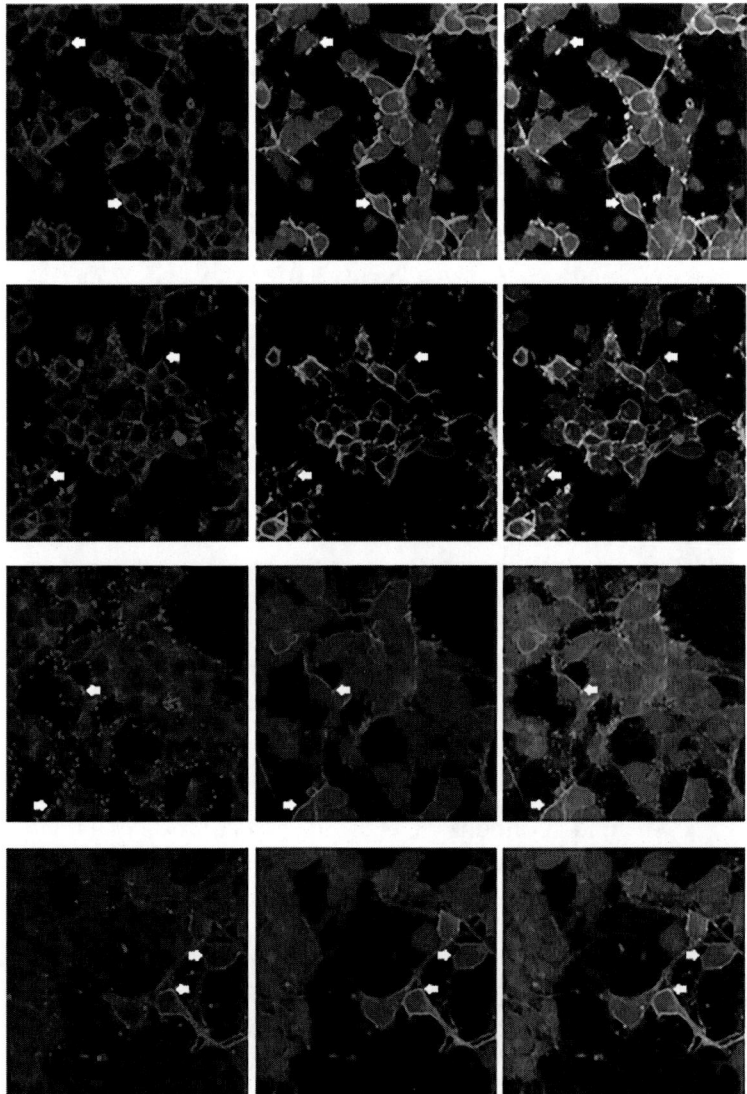

A2. **Ganglioside GT1b colocalization with protein markers related to pathology in exosomes.** Top to bottom row: Flotillin, Aβ, APP and ABCA5 (red), with GT1b (green) and DAPI (blue) staining of untreated SHSY-5Y cells. Merged images show colocalization in yellow (white arrows).

REFERENCES

Badawy S. M. M., Okada, T., Kajimoto, T., Hirase, M., Matovelo, S. A., Nakamura, S., Yoshida, D., Ijuin, T. and Nakamura, S. I. (2018). Extracellular α-synuclein drives sphingosine 1-phosphate receptor subtype 1 out of lipid rafts, leading to impaired inhibitory G-protein signaling. *J Biol Chem.*, 2018, May 25, *293*(21), 8208-8216.

Barker, R. A. and Williams-Gray, C. H. (2016) Review: The spectrum of clinical features seen with alpha synuclein pathology. *Neuropathology and applied neurobiology*, *42*, 6-19.

Beitz, J. (2014) Parkinson's disease: a review. *Front Biosci* (School Ed)., *6*, 65-74.

Black, L. and Berenbaum, M. C. (1964). Factors affecting the dye exclusion test for cell viability. *Exp. Cell Res.*, *35*, 9–13.

Botto, L., Cunati, D., Coco, S., Sesana, S., Bulbarelli, A., Biasini, E., Colombo, L., Negro, A., Chiesa, R., Masserini, M. and Palestini, P. (2014) Role of Lipid Rafts and GM1 in the Segregation and Processing of Prion Protein. *PLOS ONE*, *9*, e98344.

Brown, D. A. and London, E. (2000) Structure and Function of Sphingolipid- and Cholesterol-rich Membrane Rafts. *Journal of Biological Chemistry*, *275*, 17221-17224.

Canerina-Amaro, A., Hernandez-Abad, L. G., Ferrer, I., Quinto-Alemany, D., Mesa-Herrera, F., Ferri, C., Puertas-Avendano, R. A., Diaz, M. and Marin, R. (2017). Lipid raft ER signalosome malfunctions in menopause and Alzheimer's disease. *Frontiers in bioscience (Scholar edition)*, *9*, 111-126.

Chan, P., DeLanney, L. E., Irwin, I., Langston, J. W. and Di Monte, D. "Rapid Atp Loss Caused by 1-Methyl-4-Phenyl-1,2,3,6-Tetrahydropyridine in Mouse Brain". *J Neurochem*, *57*, no. 1, (Jul 1991), 348-51.

Chivet, M., Hemming, F., Pernet-Gallay, k., Fraboulet, S. and Sadoul, R. (2012). Emerging Role of Neuronal Exosomes in the Central Nervous System. *Frontiers in Physiology*, *3*.

Chu, Y., Goldman, J. G., Kelly, L., He, Y., Waliczek, T. and Kordower, J. H. (2014). Abnormal alpha-synuclein reduces nigral voltage-dependent anion channel 1 in sporadic and experimental Parkinson's disease. *Neurobiology of Disease*, *69*, 1-14.

Danzer, K. M., Kranich, L. R., Ruf, W. P., Cagsal-Getkin, O., Winslow, A. R., Zhu, L., Vanderburg, C. R. and McLean, P. J. (2012). Exosomal cell-to-cell transmission of alpha synuclein oligomers. *Molecular Neurodegeneration*, *7*, 42.

Diaz, M., Fabelo, N., Ferrer, I. and Marin, R. (2018). "Lipid raft aging" in the human frontal cortex during nonpathological aging: gender influences and potential implications in Alzheimer's disease. *Neurobiology of aging*, *67*, 42-52.

Edgar, J. R. (2016). QandA: What are exosomes, exactly? *BMC Biology*, *14*, 46.

Emanuele, M., Esposito, A., Camerini, S., Antonucci, F., Ferrara, S., Seghezza, S., Catelani, T., Crescenzi, M., Marotta, R., Canale, C., Matteoli, M., Menna, E. and Chieregatti, E. (2016) Exogenous Alpha-Synuclein Alters Pre- and Post-Synaptic Activity by Fragmenting Lipid Rafts. *EBioMedicine*, *7*, 191-204.

Eriksson, I., Nath, S., Bornefall, P., Giraldo, A. M. and Ollinger, K. (2017). Impact of high cholesterol in a Parkinson's disease model: Prevention of lysosomal leakage versus stimulation of alpha-synuclein aggregation. *European journal of cell biology*, *96*, 99-109.

Fabelo, N., Martin, V., Marin, R., Moreno, D., Ferrer, I. and Diaz, M. (2014). Altered lipid composition in cortical lipid rafts occurs at early stages of sporadic Alzheimer's disease and facilitates APP/BACE1 interactions. *Neurobiology of aging*, *35*, 1801-1812.

Fantini, J. (2007) Interaction of proteins with lipid rafts through glycolipid-binding domains: biochemical background and potential therapeutic applications. *Current medicinal chemistry*, **14**, 2911-2917.

Farooqui, S. M. (1994). Induction of adenylate cyclase sensitive dopamine D2-receptors in retinoic acid induced differentiated human neuroblastoma SHSY-5Y cells. *Life Sciences*, **55**, 1887-1893.

Fink, A. L. (2006). The aggregation and fibrillation of alpha-synuclein. *Accounts Chem Res*, *39*, 628–634.

Gallea, J. I., Ambroggio, E. E., Vilcaes, A. A., James, N. G., Jameson, D. M. and Celej, M. S. (Aug 24, 2018). "Amyloid Oligomerization of the Parkinson's Disease Related Protein Alpha-Synuclein Impacts on Its Curvature-Membrane Sensitivity." [In eng]. *J Neurochem.*

Grey, M., Dunning, C. J., Gaspar, R., Grey, C., Brundin, P., Sparr, E. and Linse, S. (2015). Acceleration of alpha-synuclein aggregation by exosomes. *The Journal of biological chemistry*, **290**, 2969-2982.

Kim, W. S. and Halliday, G. M. (2012) Changes in sphingomyelin level affect alpha-synuclein and ABCA5 expression. *Journal of Parkinson's disease*, **2**, 41-46.

Ledeen, R. W. and Wu, G. (2015). The multi-tasked life of GM1 ganglioside, a true factotum of nature. *Trends in biochemical sciences*, **40**, 407-418.

Lemkul, J. A. and Bevan, D. R. (2013). Aggregation of Alzheimer's amyloid beta-peptide in biological membranes: a molecular dynamics study. *Biochemistry*, **52**, 4971-4980.

Levental, I., Grzybek, M. and Simons, K. (2010) Greasing their way: lipid modifications determine protein association with membrane rafts. *Biochemistry*, **49**, 6305-6316.

Levental, I. and Veatch, S. (2016). The Continuing Mystery of Lipid Rafts. *Journal of molecular biology*, **428**, 4749-4764.

Lingwood, D. and Simons, K. (2010). Lipid rafts as a membrane-organizing principle. *Science* (New York, N.Y.), *327*, 46-50.

Marin, R. (2011) Signalosomes in the brain: relevance in the development of certain neuropathologies such as Alzheimer's disease. *Front Physiol.*, *2*, 23-30.

Marin, R., Fabelo, N., Fernandez-Echevarria, C., Canerina-Amaro, A., Rodriguez-Barreto, D., Quinto-Alemany, D., Mesa-Herrera, F. and Diaz, M. (2016) Lipid Raft Alterations in Aged-Associated Neuropathologies. *Current Alzheimer research*, *13*, 973-984.

Marin, R., Fabelo, N., Martin, V., Garcia-Esparcia, P., Ferrer, I., Quinto-Alemany, D. and Diaz, M. (2017). Anomalies occurring in lipid profiles and protein distribution in frontal cortex lipid rafts in dementia with Lewy bodies disclose neurochemical traits partially shared by Alzheimer's and Parkinson's diseases. *Neurobiology of aging*, *49*, 52-59.

Marin, R. and Diaz, M. (2018) Estrogen Interactions With Lipid Rafts Related to Neuroprotection. Impact of Brain Ageing and Menopause. *Frontiers in neuroscience*, *12*, 128.

Marini, A. M., Lipsky, R. H., Schwartz, J. P. and Kopin, I. J. (Apr 1992). "Accumulation of 1-Methyl-4-Phenyl-1,2,3,6-Tetrahydropyridine in Cultured Cerebellar Astrocytes." *J Neurochem*, *58*, no. 4, 1250-8.

Marsh, S. E. and Blurton-Jones, M. (2012). Examining the mechanisms that link β-amyloid and α-synuclein pathologies. *Alzheimer's Research and Therapy*, *4*, 11.

Martinez, Z., Zhu, M., Han, S. and Fink, A. L. (2007). GM1 specifically interacts with alpha-synuclein and inhibits fibrillation. *Biochemistry*, *46*, 1868-1877.

Meredith, G. E., Totterdell, S., Beales, M. and Meshul, C. K. (Sep 2009). "Impaired Glutamate Homeostasis and Programmed Cell Death in a Chronic Mptp Mouse Model of Parkinson's Disease". *Exp Neurol*, *219*, no. 1, 334-40. https://doi.org/10.1016/j.-expneurol.2009.06.005.

Meredith, Gloria E., Adrian, G. Dervan. and Susan, Totterdell. (2005//2005). "Activated Microglia Persist in the Substantia Nigra of a Chronic Mptp Mouse Model of Parkinson's Disease." Paper presented at the The Basal Ganglia VIII, Boston, MA.

Meredith, Gloria E. and David, J. Rademacher. (2011). "Mptp Mouse Models of Parkinson's Disease: An Update". *Journal of Parkinson's disease*, *1*, no. 1, 19-33. https://doi.org/10.3233/JPD-2011-11023.

Murphy, M. P. and LeVine, H. (2010). Alzheimer's Disease and the β-Amyloid Peptide. *Journal of Alzheimer's disease: JAD*, *19*, 311.

Nicklas, W. J., Vyas, I. and Heikkila, R. E. (Jul 1, 1985). "Inhibition of Nadh-Linked Oxidation in Brain Mitochondria by 1-Methyl-4-Phenyl-Pyridine, a Metabolite of the Neurotoxin, 1-Methyl-4-Phenyl-1,2,5,6-Tetrahydropyridine". *Life Sci*, *36*, no. 26, 2503-8.

Park, J. Y., Kim, K. S., Lee, S. B., Ryu, J. S., Chung, K. C., Choo, Y. K., Jou, I., Kim, J. and Park, S. M. (2009). On the mechanism of internalization of alpha-synuclein into microglia: roles of ganglioside GM1 and lipid raft. *Journal of neurochemistry*, *110*, 400-411.

Pirc, K. and Ulrih, N. (2011). *Alpha-Synuclein Interactions with Membranes*.

Phuyal, S., Hessvik, N. P., Skotland, T., Sandvig, K. and Llorente, A. (2014). Regulation of exosome release by glycosphingolipids and flotillins. *FEBS J.*, 2014 May, *281*(9), 2214-27.

Polanco, J. C., Li, C., Durisic, N., Sullivan, R. and Götz, J. (2018). Exosomes taken up by neurons hijack the endosomal pathway to spread to interconnected neurons. *Acta Neuropathologica Communications*, *6*.

Presgraves, S. P., Ahmed, T., Borwege, S. and Joyce, J. N. (2004) Terminally differentiated SH-SY5Y cells provide a model system for studying neuroprotective effects of dopamine agonists. *Neurotoxicity research*, *5*, 579-598.

Prusiner, S. B. (1982). Novel proteinaceous infectious particles cause scrapie. *Science (New York, N.Y.)*, *216*, 136-144.

Raposo, G. and Stoorvogel, W. (2013) Extracellular vesicles: Exosomes, microvesicles, and friends. *The Journal of Cell Biology*, **200**, 373-383.

Schneider, J. S., Cambi, F., Gollomp, S. M., Kuwabara, H., Brasic, J. R., Leiby, B., Sendek, S. and Wong, D. F. (2015). GM1 ganglioside in Parkinson's disease: Pilot study of effects on dopamine transporter binding. *Journal of the neurological sciences*, **356**, 118-123.

Schneider, J. S., Sendek, S., Daskalakis, C. and Cambi, F. (2010). GM1 ganglioside in Parkinson's disease: Results of a five year open study. *Journal of the neurological sciences*, **292**, 45-51.

Selkoe, D. J. Alzheimer's disease: genes, protein, and therapy. *Physiol. Rev.*, **81**, 741-766.

Shastry, B. S. (2003). Neurodegenerative disorders of protein aggregation. *Neurochemistry international*, **43**, 1-7.

Smilansky, A., Dangoor, L., Nakdimon, I., Ben-Hail, D., Mizrachi, D. and Shoshan-Barmatz, V. (2015). *The Voltage-dependent Anion Channel 1 Mediates Amyloid β Toxicity and Represents a Potential Target for Alzheimer Disease Therapy.*, **290**, 30670-30683.

Sweeney, P., Park, H., Baumann, M., Dunlop, J., Frydman, J., Kopito, R., McCampbell, A., Leblanc, G., Venkateswaran, A., Nurmi, A. and Hodgson, R. (2017). Protein misfolding in neurodegenerative diseases: implications and strategies. *Translational Neurodegeneration*, **6**.

Théry, C., Amigorena, S., Raposo, G. and Clayton, A. (2006). Isolation and characterization of exosomes from cell culture supernatants and biological fluids. *Curr Protoc Cell Biol.*, Chapter 3, Unit 3.22.

van Maarschalkerweerd, A., Vetri, V. and Vestergaard, B. (2015). Cholesterol facilitates interactions between alpha-synuclein oligomers and charge-neutral membranes. *FEBS letters*, **589**, 2661-2667.

Xiao, T., Zhang, W., Jiao, B., Pan, C. Z., Liu, X. and Shen, L. (2017). The role of exosomes in the pathogenesis of Alzheimer' disease. *Translational Neurodegeneration*, **6**.

Yamamoto, N., Matsubara, T., Sobue, K., Tanida, M., Kasahara, R., Naruse, K., Taniura, H., Sato, T. and Suzuki, K. (2012). Brain insulin resistance accelerates Abeta fibrillogenesis by inducing GM1 ganglioside clustering in the presynaptic membranes. *Journal of neurochemistry*, **121**, 619-628.

Yuyama, K., Yamamoto, N. and Yanagisawa, K. (2008) Accelerated release of exosome-associated GM1 ganglioside (GM1) by endocytic pathway abnormality: another putative pathway for GM1-induced amyloid fibril formation. *Journal of neurochemistry*, **105**, 217-224.

Zang, L. Y. and Misra, H. P. (1993). "Generation of Reactive Oxygen Species During the Monoamine Oxidase-Catalyzed Oxidation of the Neurotoxicant, 1-Methyl-4-Phenyl-1,2,3,6-Tetrahydropyridine." *J Biol Chem*, **268**, no. 22, 16504-12.

Zhu, M. and Fink, A. L. (2003). Lipid binding inhibits alpha-synuclein fibril formation. *The Journal of biological chemistry*, **278**, 16873-16877.

In: A Closer Look at Neurotoxicity
Editor: Gokul Krishna
ISBN: 978-1-53616-591-3
© 2020 Nova Science Publishers, Inc.

Chapter 3

NEUROTOXICITY: PERSPECTIVE IN AGE-RELATED NEURODEGENERATIVE DISEASES

Shubhangini Tiwari and Sarika Singh*, PhD
Division of Neuroscience and Ageing Biology,
Division of Toxicology and Experimental Medicine,
CSIR - Central Drug Research Institute,
Lucknow, UP, India

ABSTRACT

Neurotoxicity is the major devastating aspect that leads to onset and progression of several neurodegenerative diseases like Alzheimer's disease (AD), Parkinson's disease (PD), Amyotropic lateral sclerosis (ALS) and Huntington disease (HD) in central nervous system. Neurotoxicity implies the neurotoxin mediated neuronal death which may be environmental or endogenous but will lead to neuronal damage through apoptosis, necrosis or autophagy. Neuronal damage to specific brain region or pathway leads to precise neurodegenerative disease. Hitherto, various mechanisms have been suggested to be involved in neurotoxicity but still the lacunae exist and we are unable to understand

---

*Corresponding Author's E-mail: sarika_singh@cdri.res.in; ssj3010@gmail.com.

the cause of their initiation. Amongst all suggested mechanisms, oxidative stress and mitochondrial impairment are the prominent one and very well explored. Since the brain is foremost energy consuming organ in human body, it responds to the minutest alteration in cellular energy levels. Past research has noticeably pointed towards the declined mitochondrial functions as a known feature of aged cells and may be the key feature in age related neurodegenerative diseases. However, to claim such hypothesis we need extensive evidences, where we are presently lacking and thus unable to interpret the cause of initiation of neuronal damage. Once initiated, neuronal damage can further worsen the situation by inducing the other signaling pathways including unfolded protein responses (UPR). Protein aggregation which is the actual initiator of UPR and pathological hallmark of various neurodegenerative diseases could subsequently worsen the conditions and lead to neuronal damage. In this chapter, we discuss about the type of death and mechanisms involved in neurotoxicity which ultimately lead to neurodegenerative disorders.

INTRODUCTION

Neurotoxicity is toxicity to neurons induced by chemical, physical or biological agent and causes their death. These agents may be endogenous or environmental but significantly affect the neuronal population and causes their decreased number thus affecting physiological neuronal functions. Specific neuronal populations are responsible for the synthesis of specific neurotransmitter to execute their physiological functions like cholinergic and dopaminergic neuronal population are responsible for maintaining the levels of acetylcholine and dopamine in brain. The decreased number of these neurons, thus decreased level of these neurotransmitters causes the Alzheimer's disease (AD) (Kandimalla and Reddy, 2017) and Parkinson's disease (PD) pathology respectively, involving cognitive and motor impairment (Magrenelli et al., 2016). Such decrease in number of neurons may be due to exogenous or endogenous factors.

The exogenous environmental toxins include polyvinyl chloride (PVC), Polychlorinated, biphenyls (PCB), pesticides, insecticides as reported previously (Singh, 2017). Humans may come in contact with these toxins through inhalation, ingestion or through skin penetration which affects their mental state, the reproductive abilities and may even cause death. However, the endogenous neurotoxins such as glutamate, nitric oxide, iron and others may also play critical role, aggregated misfolded proteins being a hallmark feature of various neurodegenerative diseases that play critical role in neurotoxicity (Soto, 2003). Amyloid beta aggregation play significant role in AD pathology and alpha-synuclein aggregates are supposed to cause the dopaminergic neuronal death in PD pathology (Teresa et al., 2019). ALS is another motor neuron disease which involves the death of neurons responsible for controlling motor neuron disease, thus leading to the stiff muscles, muscle twitching and gradual weakening due to muscle loss and impaired neuronal responses.

Such patients eventually lose their ability to walk, speak, swallow, and breathe (Hobson and McDermott, 2016). Reports have suggested that environmental toxins like MPTP, rotenone play significant role in etiology of sporadic PD (Anselmi et al., 2018). These agents induced neuronal death mostly involves the oxidative stress mediated energy crisis and neuronal death which may be either apoptotic, necrotic or through autophagy. In this chapter, firstly we have briefed about the type of neuronal death, then we discuss about the various types of agents which initiate the neuronal death and finally the major mechanisms involved in neurotoxicity. We will also be discussing about the major age related neurodegenerative diseases like AD, PD, HD and ALS which occur due to neuronal loss of specific set of neuronal population (Segura-Aguilar et al., 2015). Amongst them, AD and PD hold a higher prevalence in comparison to HD and ALS (Reitz et al., 2011; Reeve et al., 2014; Beghi et al., 2006).

Types of Neuronal Cell Death Leading to Neurotoxicity

Cell death process is considered to have a physiological as well as pathological significance. Neuronal death is a common phenomenon during embryonic development, technically termed as neurogenesis (Yeo and Gautier, 2004). However, the mature and differentiated neurons do not divide and hence, the death of specific neuronal population in adults lead to specific disease such as Alzheimer's, Parkinson's, Amyotropic Lateral Sclerosis, Cerebral ischemia and trauma (Chi et al., 2018). Both neuronal death and neurodegenerative processes are suggested to be closely linked to neurogenesis (Alvarez-Buylla & Garcia-Verdugo, 2002; Gage, 2002) and mostly occurs in hypothalamus (Migaud et al., 2010). Such cell death process may involve several pathways which can be classified as apoptosis, necrosis and autophagy. All three processes are highly regulated; however the initiation causes of all are different. Apoptosis is a regular cellular mechanism which takes place to remove all the non-required cells. It has been suggested that approximately 1300 cells undergo apoptosis daily which are replaced by newly born neurons in the dentate gyrus (DG) of adult mouse, while in humans approximately 700 new neurons or 0.004% of DG neurons are added in hippocampus every day (Spalding et al., 2013). On a different note, the necrosis is initiated due to external factors in the cell or tissue like infection, toxins, or trauma. Such initiation of necrosis results in the unregulated digestion of cell components and prevents their unexpected spread and cellular functions. Autophagy is also a regulated mechanism required to disassemble the unnecessary and dysfunctional components of the cells. However, autophagy involves the orderly degradation and recycling of cellular component (Mizushima and Komatsu, 2011; Kobayashi, 2015). In the following sections, we will be discussing the details of all the processes separately.

Apoptosis

Apoptosis also called as programmed cell death was first coined by Kerr, Wyllie and Currie in 1972 (Kerr et al., 1972; Paweletz, 2001; Kerr, 2002). It is an energy dependent, principle mechanism that is activated not only during cellular stress and damage, but also governs cell destruction during normal development and morphogenesis in order to maintain the cell population in tissues (Elmore, 2017). It is the principle mechanism of cell destruction in metazoans (Elmore, 2017). Cell death cascade may be triggered by external stimuli through membrane bound receptors such as TNF-alpha (extrinsic pathway) or by internal cues involving DNA damage and p53 dependent signaling or mitochondrial signaling (intrinsic pathway). Activation of both extrinsic and intrinsic pathway leads to activation of caspases, the ultimate marker of apoptosis (Elmore, 2017). T-cell mediated cytotoxicity and perforin/granzyme pathway is an additional pathway which involves granzyme A or granzyme B for killing of cells (Martinvalet et al., 2005). Binding of Fas ligands to Fas receptors lead to binding of its adaptor protein FADD or TNF-α to TNFR1 and its adaptor cytoplasmic protein TRADD along with FADD and RIP (Hsu et al., 1995; Wajant, 2002). FADD makes complex with procaspase-8 and finally forms death-inducing signaling complex (DISC) which activates procaspase-8 through auto-catalysis (Kischkel et al., 1995). Activation of Caspase-8 leads further to the execution phase of apoptosis and thereby, cell death. Intrinsic pathway involves the mitochondrial signaling initiated in response to non-receptor mediated stimuli which leads to alteration in the inner mitochondrial membrane and decreases mitochondrial membrane potential (MMP), thereby opening up mitochondrial permeability transition (MPT) pore (Saelens et al., 2004). These changes lead to the subsequent release of two major classes of pro-apoptotic proteins from mitochondria into the cytosol. Class I includes cytochrome-c, Smac/DIABLO, and HtrA2/Omi, a serine protease (Du et al., 2000) which activates the mitochondrial associated Caspase dependent pathway. Class I protein cytochrome-c form complex with Apaf-1 and pro-caspase 9 to form apoptosome (Chinnaiyan, 1999; Hill et al., 2004). Class II proteins AIF, endonuclease G and endonuclease CAD are released from mitochondria and are translocated to the nucleus to carry out the DNA fragmentation and chromatin condensation (Joza et al., 2001); (Enari et al., 1998). Both intrinsic and extrinsic pathway end up into activation of executioner caspase-7, caspase-6 and caspase-3 which finally cleaves various substrates into apoptotic bodies that are recognized and digested by phagocytic cells (Slee et al., 2001). Major hallmarks of apoptosis include distinct change in morphology of cells, chromatin condensation and DNA fragmentation by Ca^{2+} and Mg^{2+} dependent endonucleases (Bortner et al., 1995). The target cell is transformed into apoptotic bodies which are recognized, engulfed and digested by phagocytic cells for their regulated removal. Protein cross linking is an important characteristic of apoptosis and is characterized by activation and expression of tissue transglutaminase (Nemes et al., 1996). Expression of cell surface marker in early apoptotic cell is recognized by phagocytic cell; such as movement of inner phospholipid bilayer phosphotidylserine, towards outer membrane which act as an important biochemical marker of cell undergoing apoptosis (Bratton et al., 1997).The resulting products are taken up and cleared by macrophages (Elmore, 2017). Apoptosis is also an important mechanism in various developmental processes, to get rid of pathogen invasions, healing of wound, and self-killing of autoimmune cells in central lymphoid and peripheral organs (Elmore, 2017). Oxidative stress increases with age leads to accumulation of free radicals and enhanced mitochondrial DNA damage which pathophysiologically contributes to age-induced apoptosis (Harman, 1992; Ozawa, 1995). Apoptosis is a stringently controlled process and any dysregulation may lead to various diseases including neurodegenerative diseases and cancer (Kerr et al., 1994), autoimmune diseases (Worth et al., 2006; Li et al., 1995) or developmental defects.

Necrosis

Necrosis or 'cellular accident' is an energy independent process, occuring due to sudden ATP depletion, hampering the cell survival and is caspase independent (Edinger and Thompson, 2004). Major hallmarks include vacuolation of cytoplasm, organelle swelling, plasma membrane breakdown and release of cellular contents and pro-inflammatory molecules, followed by inflammatory reactions (Edinger and Thompson, 2004). Distinct change in nuclear morphology is also observed. Proliferating cells that suffer DNA damage are also cleared through the process of necrosis through DNA repair protein Poly (ADP-ribose) polymerase (PARP) (Ha and Synder, 1999). Necrosis is triggered by PARP, as activation of PARP in response to DNA damage in proliferating cells leads to inhibition of cytoplasmic and nuclear Nicotinamide adenine dinucleotide (NAD^+) which further repress glycolysis, an important biochemical pathway for ATP production in cells through formation of pyruvate required as a substrate in oxidative phosphorylation (Ha and Synder, 1999). Therefore, cells dependent on glycolysis for energy production critically suffer with sudden ATP depletion and undergo necrosis. Necrosis or apoptosis occur depending on the type/degree of stimuli such as heat, radiation, drugs, etc. to the cells; a lower degree of damage to cells may trigger apoptosis, whereas, higher degree may stimulate necrosis (Hirsch, 1997; Zeiss, 2003). Necrosis can also be initiated by ligand binding to receptors (Vanlangenakker et al., 2008). Mostly the death receptor mediated death is due to apoptosis however, studies in few cells have shown that necrosis may also lead to death receptor mediated signaling. Reports have suggested the initiation of necrosis by death receptors including TNFR1, Fas and TRAIL (Vanlangenakker et al., 2008). It has been shown that a L929 cell, which is mostly employed to study the mechanism of TNF is capable of exhibiting the ROS induced necrosis (Fiers et al., 1999). It has also been reported that the TNFR1 and a serine – threonine kinase RIP1 could also be a crucial initiator in death receptor mediated necrosis (Festjens et al., 2007). However, studies exist suggesting the role of the caspase inhibitor in preventing the Fas-L and TNF induced necrotic death in RIP1 deficient T cells (Holler et al., 2000). Dimerization of kinase domain of RIP1 also induces the necrotic cell death (Degterev et al., 2005).

Autophagy

Macroautophagy or 'Self-cannibalization' is a cell death mechanism which involves an autophagosome, a double-membrane vesicle (Levine and Kroemer, 2008). The autophagosomes engulf the target cell and aid the digestion of cytoplasmic contents and intracellular organelles in an acidic environment, after their fusion with lysosomes, forming autolysosomes (Levine and Kroemer, 2008). Autophagosomes sometimes, fuse with endosome before finally fusing with lysosomes. Earlier, it was reported that autophagy occurred as a result of nutrient scarcity (Mizushima and Levine, 2010). However, in many neurodegenerative diseases, high rate of autophagic death depicted it as a form of "cell suicide" (Mizushima, et al., 2008). In humans, Beclin1 is the first key protein involved in formation of autophagic membrane vesicle alongwith VPS15 and VPS34 proteins, which together constitute PI3K (multiprotein) complex (Xie and Klionsky, 2007; Zeng et al., 2006). In humans, Beclin-1is encoded by the BECN1gene while in the yeast, the ortholog of Beclin-1 is autophagy-related gene 6 (Atg6) and in nematode, it is BEC-1. The Beclin 1 interacts with either BCL-2 or PI3k class III, playing a critical role in the regulation of both autophagy and cell death. Beclin1 contains the BH (BCL-2 Homology) 3 domain, which physically interacts with BCL2 and BCL-XL proteins. Such interaction of BH3 domain mediated interaction of Beclin 1 and apoptotic proteins cause inhibition of anti-apoptotic protein (BCL2, BCL-XL) which activates the pro-apoptotic proteins (BAX, BAK) (Pattingre et al., 2005). It has been reported that mutation of BH3 domain or its interacting partners lead to inhibition of autophagy (Nikoletopoulou, 2013). Both autophagy and apoptosis are interconnected mechanisms in which various autophagic proteins act as substrate in caspase- mediated apoptosis (Nikoletopoulou, 2013). It has been reported that ATG4D mediated autophagy is initiated by caspase-3 which cleaves ATG4D and in turn, ATG4D carries out the synthesis of ATG8 (LC3 in mammals), after cleavage of its C-terminus (Betin and Lane, 2009). Truncation of ATG4D leads to its translocation to mitochondria. Similarly, ATG5 (component of autophagosome) cleavage by caspase increases its translocation to mitochondria and thus, its participation in mitochondrial apoptotic signaling cascade (Nikoletopoulou, 2013). FLIP (Flice inhibitory protein) is as an inhibitor of TNF or NGF mediated apoptosis (Lee et al., 2009). FLIP binds with autophagosome elongation mediator ATG3 and prevents its binding with LC3, thereby, inhibiting autophagy (Lee et al., 2009). However, under stressed conditions, FLIP levels decrease to permit the LC3-ATG3 binding to stimulate the autophagy (Lee et al., 2009). DAPK, a calcium/calmodulin dependent kinase is activated during ER stress, and is responsible for carrying out both autophagy and apoptosis (Lerner and Kimchi, 2012). It has been observed that DAPK knockout in mouse kidney toxicity models and invitro fibroblasts lead to the reduction in ER stress induced autophagy and apoptosis (Lerner and Kimchi, 2012). It has also been suggested that DAPK phosphorylates beclin1 (BH3 domain) on Thr119 and dissociates its interaction with anti-apoptotic BCL2, thereby inducing autophagy (Nikoletopoulou, 2013).

NEUROTOXINS INDUCED NEUROTOXICITY

As described earlier, neurotoxicity is any adverse effect to neurons which lead to neuronal death. It may be due to endogenous or due to exogenous neurotoxicants. Exogenous neurotoxicants initiate the neurotoxicity when a person gets into contact with them; however, we have reviewed neurodegeneration caused due to the exogenous neurotoxicants in our previous review (Singh, 2017).In this chapter, we focus on the endogenous neurotoxins; since endogenous neurotoxicants are produced within the body and severely affect our homeostatic mechanisms. Such endogenous neurotoxicants majorly include glutamate, nitric oxide and iron. All three are well-known to play a significant physiological role, however studies have suggested their pathological role as well. Both glutamate and nitric oxide play a major role as the neurotransmitter; while iron is a crucial metal in various physiological and pathological reactions especially in the Fenton's reaction. In the following section, we shall be discussing about the toxicity exerted by these three neurotoxicants.

Glutamate Toxicity

Glutamate is an endogenous excitatory neurotransmitter in CNS and excess of L-glutamate mediated synaptic transmission called glutamate excitotoxicity negatively affects the neuronal function and causes neurodegeneration (Choi, 1988; Guan et al., 2015; Shah et al., 2016). The delicate equilibrium of glutamate is maintained by glutamate transporters (Wang et al., 2018). Somehow, if these transporters become non-functional, due to mutation or any other reason, then the glutamate concentration would increase across the glutamate receptors and cause the opening up of the ion channels, allowing calcium to enter the cell in order to cause the excitotoxicity. Studies have suggested that N-methyl-D-aspartate receptor (NMDAR) mediated excitotoxicity is dependent on the route of calcium influx, and not the calcium load, as even lower amount of calcium influx through NMDAR caused more neuronal death than much higher amount of influx through other calcium permeable channels (Zhou et al., 2013; Carvajal et al., 2016). Glutamate mostly interacts with NMDA receptors, to offer the increase in the release of calcium into the cell (Carvajal et al., 2016). Such augmented level of calcium, increases its load on mitochondria (Carvajal et al., 2016) as well as might cause the altered ER homeostasis (Ruiz et al., 2009).Stress on mitochondria may initiate the increased generation of reactive oxygen species (ROS) with which the cell needs to deal with, utilizing the support of existing antioxidative enzymes (Zorov et al., 2014). However, the depleted antioxidants level aggravates oxidative stress which further worsens the conditions. Such oxidative conditions and depleted level of antioxidative enzymes are prominently observed in neurodegenerative pathologies including AD, PD, HD and ALS (Lau and Taimianski, 2010; Cenini et al., 2019). Augmented levels of calcium, potentially increases the expression of NOS, thereby increase in the level of NO which could further interact with NMDARs and worsen the adverse conditions in cell. In support to this, it has been observed that the application of NOS inhibitor potentially attenuates the glutamate induced excitotoxicity (Bharadwaj et al., 1997).

Nitric Oxide

Nitric oxide is an unconventional endogenous biological messenger and has received the attention of a potent neurotransmitter in the brain (Boulu et al., 1994; Paul and Ekambaram, 2011). As we discussed above, the imbalance in glutamate neurotransmission results in calcium influx which binds calmodulin and activates the neuronal nitric oxide synthase (nNOS), which further convert the L-arginine to citrulline and nitric oxide (NO). Though NO acts as a neurotransmitter in brain, its excess level is neurotoxic (Džoljić et al., 2015). It has been reported that NO induced neurotoxic effects involves the glutamate induced toxic effects (Bashkatova and Rayevsky, 1998). In glutaminergic pathology also the role of NO has been reported (Raevskii et al, 2000), which is further reported to be well involved in pathologies of AD, PD,

HD and ALS (Hynd et al., 2004; Ambrosi et al., 2014; Estrada Sanchez et al., 2008; Shaw and Ince, 1997). In addition, the adverse effects of NO in cognition, epilepsy, stress, diabetes have also been reported (Paul and Ekambaram, 2011). NO ability to augment the cyclic GMP suggests its implication in the metabolic pathways of brain (Bolaños et al., 1997). NO could also execute its interaction with mitochondrial electron transport chain complexes (Brown, 1999). Most of the neurodegenerative conditions involve impaired mitochondrial complexes activities, which lead to increase in level of superoxide (Johri and Beal, 2012). Such increased level of superoxide further react with NO, to form the peroxynitrite and worsen the conditions which are already implicated in neurodegenerative diseases (Bolaños et al., 1997). Brain glial cells also express nitric oxide synthase during neurodegenerative conditions; the increased glial reactivity is a well-accepted pathological hallmark (Lee and MacLean, 2015). In this regard, the research has been reviewed elsewhere by us as well as others suggesting the implication of glial NO in neurodegenerative diseases (Yuste et al., 2015; Singh et al., 2011)).Under neurodegenerative conditions, the iNOS expression could be increased in glial cells due to inflammatory reactions, however, the concomitant activation of NADPH oxidase, could further induce the neuronal death (Brown, 2010). In the MPP+ mouse model of PD, nNOS null mice were found to be resistant towards striatal dopamine depletion (Itzhak et al., 1998). Similarly, inhibition of NO protects the brain cells against oxygen toxicity, inflammatory disorders of brain and amyloid-β aggregation in Alzheimer's disease (Dawson and Dawson, 1996).

Iron

Deregulation in the metabolism and accumulation of transition metals such as iron, zinc or copper has been linked to various neurodegenerative diseases such as Alzheimer's and Parkinson's disease (Uttara et al., 2009). Homeostasis of metal ion is maintained through the regulation of its uptake, metabolism, storage and release (Nelson, 1999). Under normal conditions, metal ions are bound to their ligands and are transported in and out of each cellular compartment through specific transporters (Salvador et al., 2010). Metal ions are required in the process of energy metabolism, as they serve as a cofactor in the transport of oxygen. However, metals not bound to their ligands and existing as free species, lead to their accumulation and generation of ROS species due to their tendency to accept and donate electrons (Halliwell, 2009). Free iron accumulation in brain precedes its abnormal interaction with various biomolecules such as lipids, proteins or nucleic acids and lead to neurodegenerative disease (Zecca et al., 2004; Barnham et al., 2014). It has also been reported that intracellular free iron pool in the CNS is known to cause both idiopathic and familial motor neuron disease (Salvador et al., 2011). Also abnormal concentrations of iron have been reported in the neurons, axons and glial cells of patients suffering with multiple sclerosis (Levine, 1997). Increase in iron concentration, leads to its accumulation in the substantia nigral neurons of PD patients and is an active substance in the lewy neurites (Kell, 2010). Iron is taken up by body cells through nutrient rich food and is bound to protein ferritin, a storage protein. Its uptake mainly takes place in the gastrointestinal tract, from where it is transported in the blood especially through transferrin receptors present on enterocytes. Since, transferrin bound iron cannot cross blood brain barrier (BBB), it is recognized by the transferrin receptors on the endothelial cells of the brain, followed by its endocytosis. It is also taken up by neurons through divalent metal transporter-1 (DMT1) or transferrin receptors (Salvador et al., 2011). Iron can be stored in brain cells either as ferritin (Bartzokis, et al., 2007) or may interact with other reactive neuromelanin or hemosiderin (Zucca et al., 2015). Reduction of molecular oxygen in mitochondria leads to production of hydrogen peroxide, which reacts with free iron through Haber-Wiess reaction to form highly reactive hydroxyl species (Lewen et al., 2000). These hydroxyl species are toxic to cells and damage the brain cells highly implicated in age related neurodegenerative diseases (Lewen et al., 2000).

MAJOR SIGNALING PATHWAY OF NEURONAL SURVIVAL AND DEMISE

In brain, the neuronal viability is attained by complex cellular interactions. These may be between neuron-neuron or neuron-glia. Most prominent interaction between astrocytes and neuron is maintenance of GSH levels in the neurons (Dringen et al., 1999; Dringen and Hirrlinger, 2003), which is known for more than a past decade. Another resident glial cell, microglia also functionally interact with neurons and play significant role in cellular network of brain (Bruce-Keller, 1999; Rossi and Volterra, 2009; Eyo and Wu, 2013). Microglial cells regulate innate immunity and contribute in the adaptive immune responses of the CNS. These cells exert both neuroprotective as well as the neurodegenerative effects as reviewed by us earlier (Singh et al., 2011). Recently, the critical bidirectional interactions of microglia have been suggested which play substantial role in brain physiology as well as pathology (Pósfai et al., 2019). Such bidirectional interactions are among neurons and are also between neurons and astrocytes. These essential cellular interactions could be affected by any kind of cellular stress including neurodegenerative disease, stroke, epilepsy, migraine, and neuroinflammatory diseases (Ricci et al., 2009). Under physiological conditions, these interactions remain tightly regulated and exert only beneficial effects to neurons however; under stress such responses initially exert the neuroprotective effects but prolonged stress lead to neurodegenerative reactions. Such cellular stresses disturb the existing equilibrium between neuronal survival and the cell death signaling, severely impairing the neuronal functions.

These degenerative events include various deaths signaling pathway specifically related to energy deprivation. As said above, the neuron glia interaction is metabolically paired, disturbance in this interaction primarily relates to the energy deprivation. This deprivation is related to impaired mitochondrial activity, which further initiates several signaling pathways including oxidative stress, endoplasmic reticulum stress and apoptosis (Goswami et al., 2017). In the following section, we will be describing the major cellular events related to neuronal death.

Oxidative Stress and Mitochondrial Dysfunction

Oxidative stress is a consequence of imbalance between generation of reactive oxygen species and their neutralization from anti oxidative enzymes. Oxidative stress constitutes the oxidation of lipids, proteins and DNA and involves the various reactive species including nitric oxide, hydrogen peroxide and reactive hydroxyl radicals (Ferrari, 2000). NO and superoxide can react with each other to form more toxic species, peroxynitrite (Radi, 2018). Brain physiology requires a high amount of metal ions like copper, zinc and iron. These trace metals mostly act as cofactor of enzymes involved in various physiological reactions including electron transport, oxygen transportation, proteins modification, neurotransmitter synthesis, redox reactions, immune responses, cell adhesion and in metabolism of proteins and carbohydrate (Chen et al., 2016). These metals accumulate in the brain and their deficiency is related to various neurological diseases (Chen et al., 2016). Stress or injury induced augmented level of ROS could initiate various neurodegenerative reactions including modifications in macromolecules of cells (Circu and Aw, 2010; Butterfield et al., 1998). Oxidative stress also induces the calcium influxes which elicit aberrant signaling pathways and initiates the apoptotic mechanisms. Calcium influx induces a major excitotoxic responses involving glutamate receptor as implicated in several neurological pathologies (Dong et al., 2009). Mitochondrion, a powerhouse of the cell is particularly sensitive to changes in energy metabolism (due to altered electron transport chain), intracellular calcium levels, neuronal depolarization or excitotoxic responses in cell (Dong et al., 2009). Mitoptosis or "mitochondrial suicide" occurs with increased ROS production and mitochondrial dysfunction, which leads to increase in mitochondrial outer membrane permeabilization (MOMP) and results in mitochondrial membrane potential loss (Jangamreddy and Los, 2012). Increase in MOMP, a BAK/BAX dependent mechanism leads to interaction of DDP/TIMM8a (mitochondrial intermembrane space protein) with DRP1 after being released into the cytoplasm. DRP1 increases the process of mitochondrial fission during cellular stress and participates in mitoptosis; whereby, end products or mitotoptic bodies via atypical exocytosis are released into the extracellular space (Arnoult et al., 2005). Poor capacity of the brain to tolerate oxidative stress and mitochondrial dysfunction and its meager regenerative capacity, leads to activation of downstream apoptotic cascade and autophagy-lysosomal pathway that is eventually responsible for neuronal death (Barnham, 2004).

ER Stress and Unfolded Protein Response

Endoplasmic reticulum (ER) is a cell organelle that carries out synthesis of nascent proteins and transports them to different locations (Lindholm et al., 2006). It is responsible for correct folding and post-translation modification of these proteins so that they may carry out normal cell function. Stress or injury mediated functional ER impairment will lead to improper folding and transport resulting in their accumulation (Paschen et al., 2001; Breckenridge et al., 2003; Rao et al., 2004). Several environmental or endogenous factors such as exposure to metal ions, oxidative stress and prior accumulation of misfolded proteins, lead to destabilization in the conformation of protein due to mutations induced in the native monomeric proteins and cause formation of fibrillar aggregated form of unstable proteins such as β-amyloid, Aβ (Soto, 2003). Such accumulated unfolded or misfolded proteins lead to ER stress which ultimately initiates the unfolded protein responses (UPRs) to achieve homeostasis (Lindholm et al., 2006). The UPR modifies the native expression and function of certain proteins and arrest the translation to prevent the accumulation of proteins (Malhotra and Kaufman, 2007)). Since ER is rich in calcium dependent molecular chaperones such as GRP78, GRP94 and calreticulin, any alterations in calcium homeostasis triggers the ER stress and ER assisted degradation (ERAD) of proteins (Paschen et al., 2001; Breckenridge et al., 2003; Rao et al., 2004), which leads to activation of various apoptotic and cell death pathway such as activation of mitogen-activated protein kinases (MAPKs), Jun N-terminal kinase (JNK) and transcription factor nuclear factor-κB (NF-κB) (Kouroku et al., 2007; Adolph et al., 2013). The alteration in UPR also leads to activation of autophagy (Senft and Ronai, 2015). ER membrane located transmembrane kinase IRE1 could activate the MAPks and adaptor TRAF2 which leads to degradation of HTT, mutant Huntingtin protein (Fouillet et al., 2012). Similarly, the altered levels of ATF4 lead to activation of ATG12, ATG5 and Beclin1 expression (B'Chir et al., 2013). Although, the cellular signaling associated with ER stress and cell death is still under investigation and the dysfunction in Ubiquitin Proteosome System (UPS) is presumed to be the most important contributor and propagator of ER stress (Imai et al., 2001). The activity of UPS is reduced during misfolded protein aggregation or high oxidative stress conditions which affects neuronal function and causes neurodegeneration (Imai et al., 2001; Korhonen et al., 2004; Bence et al., 2001). For most major neurodegenerative diseases, the accumulation and aggregation of misfolded, unfolded or mutant proteins is a common pathological hallmark which hamper the normal structure and function of the neuron and the process of synapse, leading to disruption in neuronal connectivity, thereby, causing cell death (Soto, 2003). Report has been suggested that the function of UPS is found to be compromised either due to the accumulation of misfolded proteins or due to accelerated oxidative stress and apoptotic cellular signaling (Bence et al., 2001). Mitigation in UPS activity also activates the positive feed-forward cycle of accumulation of proteins and exasperation of disease. Accumulation of neurotoxins such as Aβ plaques and neurofibrillary tangles (NFT), an intracellular filamentous structure of tau protein in Alzheimer's disease brain is also linked to hyper activation of PERK- EIF2α pathway and over expression of GRP78 & transcription factor CHOP (Hoozemans et al., 2005; Selkoe, 2001), suggesting role of ER stress in AD pathology (Ross and Poirier, 2004). Similarly, gain of function of alpha-synuclein found in lewy neurites of Parkinson's disease brain suppresses the UPS (Greenamyre and Hastings, 2004). Compounds such as 6- hydroxy dopamine, N-methyl-4-phenyl-1, 2,3,6-tetrahydroyridine or its active derivative MPP$^+$, rotenone are reported to induce the oxidative stress, mitochondrial dysfunction and ER stress with over expression of the molecular chaperone GRP78, transcription factor CHOP along with phosphorylation of the ER stress kinases, IRE1 and PERK. ER stress leads to activation of mouse caspase-12 and its human homologue caspase-4 which ultimately lead to activation of caspase-3 and neuronal death (Morishima et al., 2004). Studies on ER stress and its associated signaling pathway needs further studies and a deeper understanding for decoding the therapeutic targets of neurodegeneration.

p53 as a Major Mediator of Apoptosis, Autophagy and Neurodegeneration

DNA damage and repair, cellular ageing and senescence, apoptosis, hypoxia, uncontrolled oncogene expression stimulating cell-cycle checkpoints, calcium overload are few key factors which play significant role in activation of p53 protein. Reports have suggested the increasein level of p53 during neurodegenerative conditions (Morrison and Kinoshita, 2000). p53 induces both extrinsic as well as intrinsic apoptotic signaling cascade and has been found to interact with BCL2, suppressing its expression or transcriptional activation of BCL2 multidomain family protein, BAX and other BH3-only proteins PUMA and BID to induce apoptotic signaling (Schuler and Green, 2001). p53 can also directly interact with apoptosome member, APAF1, caspase-9 and effector caspase-6. In addition to intrinsic pathway, it interacts with extrinsic death pathway proteins and activates apoptotic machinery (Fridman and Lowe, 2003). Therefore, p53 has been studied in both activation and repression of apoptosis. Autophagy through p53 takes place via AMP kinase (inhibition of mTOR) and lysosomal protein, DRAM (damage-regulated autophagy modulator) transactivation, respectively (Crighton et al., 2006). Genotoxic stress is the major reason for DRAM induced autophagy by p53.

Recent findings suggested that p53 block autophagy and simultaneously increase apoptosis in cytoplasm of the cell (Mrakovcic and Fohlich, 2018). p53 inhibition in cells is known to induce autophagy which serves as a cytoprotective measure of cell during hypoxia or nutrient depletion. It was observed that cytoplasmic, and not nuclear, p53 could potentially act on p53 deficient cells and inhibit autophagy, thereby suggesting the role of p53 depending on its subcellular location (Nikoletopoulou et al., 2013). In neurodegenerative diseases such as AD and PD, the expression of p53 substantially increases in neurons, leading to activation of various apoptotic and autophagic pathways, henceforth, neuronal death (Chang et al., 2011). Preclinical studies support p53 inhibition as an important therapeutics in targeting neurological conditions (Ranganathan and Bowser, 2010; Jabelli et al., 2011).

EVIDENCING FOR NEUROTOXICITY IN AGE RELATED NEURODEGENERATIVE DISEASES

Neurotoxicity in Central Nervous System is known to induce through oxidative stress, mitochondrial dysfunction and energy crisis, ER stress and ERAD in the neuronal cells. Perpetual occurrence of neurotoxic mechanisms leads to onset on various neurodegenerative disease. In the following section, we discuss about the evidences, suggesting role of discovered major neurotoxic mechanisms involved in major neurodegenerative diseases.

Evidences in AD

It is a neurological disorder associated with the loss of cognitive functions. Known etiology is significant due to degeneration of cholinergic neurons specifically located in cortical and hippocampal region of the brain. Reports have suggested that such neurotoxicity majorly occurs due to accumulation of $A\beta$ plaques and filamentous tau protein named as NFT (Soto, 2003; Nisbet et al., 2014; Kametani and Hasegawa, 2018; Simic et al., 2016). Mutations in protein amyloid precursor protein (APP) and presinilins have been observed in familial type of AD (Mohmmad Abdul et al., 2006). These mutations lead to increased production of toxic $A\beta$, which further causes toxic plaque formation as observed in the brain of AD patients (Mohmmad Abdul et al., 2006). Sporadic AD also involves alteration in APP processing and eventually formation of amyloid plaques and disturbance in calcium metabolism (Zhang et al., 2011). Accumulation of $A\beta$ causes increased oxidative stress due to ROS production, ER stress and microglial activation (Uttara et al., 2009). γ-secretase protein complex consists of presinilin-1 (PS1) and presinilin-2 (PS2) which participate in APP processing (Strooper et al., 2012). It was observed that mutation in presinilins was associated with accumulation of amyloid, increased

ROS production, disruption in calcium homeostasis and susceptibility of neurons towards excitotoxic injury (Gliechmann and Mattson, 2011; Lewarenz and Maher, 2015). Mutant PS1 disrupts the UPRs and inhibit the ER membrane located transmembrane kinase, IRE1 (Doyle et al., 2011). It also increases the caspase activity and lead to neuronal death as reported in the human neuroglioma cells (Kovacs et al., 1999) and mice (Chan et al., 2002).

Evidences in PD

It is one of the most common movement neurodegenerative disorders, caused due to death of dopaminergic neurons of substantia nigra and their terminals in striatum of brain which affect approx. 1 percent of the aged population (Lang and Lozano, 1998; Reeve et al., 2014). It is characterized by bradykinesia, tremor and rigidity (Lang and Lozano, 1998). Neurotoxic mechanisms are induced by both environmental and internal cues. One of the most common neurotoxic mechanisms observed during disease pathology is increased oxidative stress and dysfunctional respiratory chain which leads to the loss of nigrostriatal neurons (Reeve et al., 2014). Although dopamine is the major neurotransmitter of the nigrostriatal neurons, increase in dopamine turnover catalysed via monoamine oxidase (MAO) leads to neurotoxicity through generation of oxidizing species such as H_2O_2 and oxidative stress in neuronal cells (Breese and Traylor, 1971). Auto-oxidation of dopamine also leads to generation of H_2O_2 and quinines, further contributing to the oxidative stress and neuronal death (Seitz et al., 2000). Alpha-synuclein, a pathological hallmark of disease is encoded by SNCA gene, and mutations such as A53T, A30P or E46K or triplication of the SNCA locus was associated with the gain of function and accumulation of alpha-syn in lewy neuritis (a neuropathological condition) causing neurotoxicity due to mutation in the SNCA gene of Parkinson's disease patients (Stefanis, 2012).

Environmental toxins mainly include MPTP and its metabolite MPP^+, rotenone, 6-OHDA and methamphetamine leading to occurrence of sporadic PD (Stefanis, 2012). MPP^+ and rotenone both inhibit the complex I activity of mitochondria and energy metabolism pathway leading to mitochondrial dysfunction and Parkinson's disease. Studies have shown that exposure to rotenone or MPP^+ or methamphetamine in mouse or non-primate models leads to decreased level of dopamine in SN and striata regions of brain characterized by the down regulation of rate limiting enzyme tyrosine hydroxylase utilized in dopamine synthesis. In normal brain, concentration of neuromelanin increases with age, however, in PD, it seems to be lost suggesting decrease in level of neuroprotective activity in substantia nigra. MPP^+ and rotenone both promote increase in levels of alpha-syn promoting its accumulation in neurons and increased ROS, mitochondrial dysfunction and ER stress, causing neurodegeneration in dopaminergic neurons (Zeng et al., 2018). Lipid peroxidation is also an essential marker of oxidative stress in PD pathology which occurs due to the unsaturated lipids having high susceptibility to oxidative modification. Lipid peroxidation, leads to generation of lipid peroxy radicals which are highly reactive compounds produced as a result of radical attack on double bonds of unsaturated fatty acids (Dexter, et al., 1989). This mechanism initiates a chain reaction, which leads to generation of various breakdown metabolites such as acrolein, malondialdehyde, 4-hydroxy- 2, 3-nonenal (HNE). These metabolic products are found to be increased in patients suffering from PD, thus highlighting the role of oxidative stress in diseased conditions (Barhnam, et al., 2004).

Evidences in ALS

Amyotrophic lateral sclerosis (ALS) or motor neuron disease is characterized as the loss of neurons in the spinal cord, brain stem and cortex (Rowland and Shneider, 2001). Classical hallmarks include paralysis and atrophy of muscles. Disease is mainly caused due to increase in oxidative stress, altered calcium signaling, hypoxia, accumulation of toxic products and excitotoxicity (Turner et al., 2005). Familial ALS is caused due to mutation in enzyme responsible for quenching ROS, SOD1 (Cu/Zn superoxide dismutase) (Soto, 2003; Turner et al., 2005). Mutation in SOD1 gene is associated with increase in activation of caspases and neuronal death in motor neurons (Pasinelli et al., 1999). Mutant SOD1 is known to interact with anti-apoptotic Bcl2 and is known to form aggregates in mitochondria, promoting mitochondrial dysfunction and changes in energy metabolism (Pasinelli et al., 2004). Aggregates of intracellular cytoplasmic inclusions in the spinal cord of ALS induced mice, showed ubiquitin protein such as Dorfin, an E3 ubiquitin ligase responsible for degradation of SOD1 (Hishikawa et al., 2003; Sone et al., 2010). The function of UPS in diseased motor neuron has been observed to be severely compromised, as a result, glutamate induced neurotoxicity becomes evident with significant increase in ROS, intracellular calcium levels which profoundly damage the mitochondria and ER and eventually cause neurodegeneration (Wootz et al.,2004; Nakagawa and Yuan, 2000).

Evidences in HD

Polyglutamine (polyQ) rich repeats in different proteins have been observed in Huntington's disease (Hinz et al., 2012). Accumulation of such proteins with the polyQ expansions causes altered gene expression, disturbed protein- protein interaction, dysfunctional mitochondria and up-regulated UPS activity in the affected neurons (Cortes and Spada, 2015; Matilla-Duenas et al., 2014). Huntington protein affects the calcium signaling in ER of the neuron through sensitization of its IP3 receptors and initiate functional impairment of the ER (Lipinski and Yuan, 2004; Tang et al., 2003).

PolyQ expansions co-localize with ER stress induced chaperones like HSP70 and HSP40 (Lipinski and Yuan, 2004). Compromised UPS and proteosome have been found in neurons with mutations in SCA3 gene which lead to activation of IRE1 and PERK along with, up regulated GRP78 and GADD153 (Nishitoh et al., 2002). Mutant huntingtin protein is also known to interact with mitochondrial DRP-1(Shirendeb et al., 2011), increasing its GTPase activity and mitochondrial fragmentation, thereby, decreasing the rate of its axonal transport and distribution causing synaptic degeneration (Reddy et al., 2011). Impaired mitochondrial functions affect the levels of ATP in nerve terminals and henceforth, lead to onset of neurodegenerative disease and neuronal death.

CONCLUSION AND FUTURE PERSPECTIVES

In this chapter, we have laid emphasis on the complex relationship between neurotoxicity associated with major neurodegenerative disorders. Both endogenous and the environmental neurotoxins, activate various cell death pathways and lead to onset of several neurodegenerative diseases. Substantial evidences shed light on the cross-talk amongst cellular organelle like mitochondria, endoplasmic reticulum and lysosomal-autophagy pathway which functionally impair and accelerate the neurodegenerative processes that consequently lead to the neuronal death.

Despite intense research on neurotoxicity, its causes and mechanisms, the intracellular pathways of neuronal death are still under investigation. Better understanding of neurotoxicity and categorical investigation of its associated mechanisms in future, would facilitate in developing better approach and therapeutics towards prevention of neuronal loss during ageing and neurological diseases.

REFERENCES

Adolph, T.E., Tomczak, M.F., Niederreiter, L., Ko, H.J., Böck, J., Martinez-Naves, E., Glickman, J.N., Tschurtschenthaler, M., Hartwig, J., Hosomi, S., Flak, M.B., Cusick, J.L., Kohno, K., Iwawaki, T., Billmann-Born, S., Raine, T., Bharti, R., Lucius, R., Kweon, M.N., Marciniak, S.J., Choi, A., Hagen, S.J., Schreiber, S., Rosenstiel, P., Kaser, A., Blumberg, R.S. (2013). Paneth cells as a site of origin for intestinal inflammation.*Nature,* 503(7475): 272-276.

Alvarez-Buylla, A., Garcia-Verdugo, J.M. (2002). Neurogenesis in adult subventricular zone. *J Neurosci,* 22: 629-634.

Ambrosi, G., Cerri, S., Blandini, F. (2014). A further update on the role of excitotoxicity in the pathogenesis of Parkinson's disease. *J Neural Transm* (Vienna), 121(8): 849-859.

Anselmi, L., Bove, C., Coleman, F.H., Le, K., Subramanian, M.P., Venkiteswaran, K., Subramanian, T., Travagli, R.A. (2018). Ingestion of subthreshold doses of environmental toxins induces ascending Parkinsonism in the rat. L. *npj Parkinson's Disease,* volume 4, Article number: 30.

Arnoult, D., Rismanchi, N., Grodet, A., Roberts, R.G., Seeburg, D.P., Estaquier, J., Sheng, M., Blackstone, C. (2005). Bax/Bak-dependent release of DDP/TIMM8a promotes Drp1-mediated mitochondrial fission and mitoptosis during programmed cell death. *Curr Biol,* 15: 2112–2118.

Arrasate, M., Finkbeiner, S. (2012). Protein aggregates in Huntington's disease. *Exp Neurol,* 238(1): 1-11.

Barnham, K.J., Masters, C.L., Bush, A.I. (2004). Neurodegenerative diseases and oxidative stress. *Nature reviews | drug discovery,* 3: 205- 214.

Bartzokis, G., Tishler, T.A., Lu, P.H., Villablanca, P., Altshuler, L.L., Carter, M., Huang, D., Edwards, N., Mintz, J. (2007). Brain ferritin iron may influence age-and gender-related risks of neurodegeneration. *Neurobiol Aging,* 28(3): 414-423.

Bashkatova, V.G.,Rayevsky, K.S. (1998). Nitric oxide in mechanisms of brain damage induced by neurotoxic effect of glutamate. *Biochemistry* (Mosc), 63(7): 866-873.

B'chir, W., Maurin, A.C., Carraro, V., Averous, J., Jousse, C., Muranishi, Y., Parry, L., Stepien, G., Fafournoux, P., Bruhat, A. (2013). The eIF2α/ATF4 pathway is essential for stress-induced autophagy gene expression. *Nucleic Acids Res,* 41(16): 7683-7699.

Beghi, E., Logroscino, G., Chiò, A., Hardiman, O., Mitchell, D., Swingler, R., Trayno, B.J., EURALS Consortium. (2006). The epidemiology of ALS and the role of population-based registries. *Biochim Biophys Acta,* 1762(11-12): 1150-1157.

Bence, N.F., Sampat, R.M., Kopito, R.R. (2001). Impairment of the ubiquitin proteasome system by protein aggregation. *Science,* 292: 1552–1555.

Betin, V.M., Lane, J.D., Caspase cleavage of Atg4D stimulates GABARAP-L1 processing and triggers mitochondrial targeting and apoptosis, *J. Cell Sci,* 122: 2554–2566.

Bhardwaj, A., Northington, F.J., Ichord, R.N., Hanley, D.F., Traystman, R.J., Koehler, R.C. (1997). Characterization of Ionotropic Glutamate Receptor– Mediated Nitric Oxide Production In-Vivo in Rats. *Stroke,* 28: 850-857.

Blokhuis, A.M., Groen, E.J., Koppers, M., van den Berg, L.H., Pasterkamp, R.J. (2013). Protein aggregation in amyotrophic lateral sclerosis. *Acta Neuropathol,* 125(6): 777-794.

Bolaños, J.P., Almeida, A., Stewart, V., Peuchen, S., Land, J.M., Clark, J.B., Heales, S.J. (1997). Nitric oxide-mediated mitochondrial damage in the brain: mechanisms and implications for neurodegenerative diseases. *J Neurochem,* 68(6): 2227-2240.

Bortner, C.D.,Oldenburg, N.B.,Cidlowski, J.A. (1995). The role of DNA fragmentation in apoptosis. *Trends Cell Biol,* 5(1): 21-26.

Boulu, R.G.,Plotkine, M.,Buisson, A. (1994). Nitric monoxide, a new neurotransmitter in the central nervous system. *Ann Pharm Fr,* 52(2): 69-80.

Bratton, D.L., Fadok, V.A., Richter, D.A., Kailey, J.M., Guthrie, L.A., Henson, P.M. (1997). Appearance of phosphatidylserine on apoptotic cells requires calcium-mediated nonspecific flip-flop and is enhanced by loss of the aminophospholipid translocase. *J Biol Chem,* 272(42): 26159-26165.

Breckenridge, D.G., Germain, M., Mathai, J.P., Nguyen, M. Shore, G,C, (2003). Regulation of apoptosis by endoplasmic reticulum pathways. *Oncogene,* 22: 8608–8618.

Breese, G.R., Traylor, T.D. (1971). Depletion of brain noradrenaline and dopamine by 6-hydroxydopamine. *Br J Pharmacol,* 42(1): 88–99.

Brown, G.C. (1999). Nitric oxideand mitochondrial respiration. *Biochim Biophys Acta.*1411(2-3): 351-369.

Brown, G.C. (2010). Nitric oxideand neuronal death. *Nitric Oxide,* 23(3): 153-165.

Bruce-Keller, A.J. (1999). Microglial-neuronal interactions in synaptic damage and recovery. *J Neurosci Res,* 58(1): 191-201.

Butterfield, D.A., Koppal, T., Howard, B., Subramaniam, R., Hall, N., Hensley, K., Yatin, S., Allen, K., Aksenov, M., Aksenova, M., Carney, J. (1998). Structural and functional changes in proteins induced by free radical-mediated oxidative stress and protective action of the antioxidants N-tert-butyl-alpha-phenyl nitrone and vitamin E. *Ann N Y Acad Sci,* 854: 448-462.

Carvajal, F.J., Mattison, H.A., Cerpa, W. (2016). Role of NMDA Receptor-Mediated Glutamatergic Signaling in Chronic and Acute Neuropathologies. *Neural Plast,* 2016: 2701526.

Carvajal, F.J., Mattison, H.A., Cerpa, W. (2016). Role of NMDA Receptor-Mediated Glutamatergic Signaling in Chronic and Acute Neuropathologies. *Neural Plast,* 2701526.

Cenini, G., Lloret, A., Cascella, R. (2019). "Oxidative Stress in Neurodegenerative Diseases: From a Mitochondrial Point of View," *Oxidative Medicine and Cellular Longevity,* vol. 2019, Article ID 2105607, 18 pages.

Chan, S.L.,Culmsee, C.,Haughey, N.,Klapper, W.,Mattson, M.P. (2002). Presenilin-1 mutations sensitize neurons to DNA damage-induced death by a mechanism involving perturbed calcium homeostasis and activation of calpains and caspase-12. *Neurobiol Dis,* 11(1): 2-19.

Chang, J.R., Ghafouri, M., Mukerjee, R., Bagashev, A., Chabrashvili, T., Sawaya, B.E. (2011). Role of p53 in neurodegenerative diseases. *Neurodegener Dis,* 9(2): 68–80.

Chen, P., Miah, M.R., Aschner, M. (2016). *Metals and Neurodegeneration.* F1000Res, 5: F1000 Faculty Rev-366.

Chi, H., Chang, H.Y., Sang, T.K. (2018). Neuronal Cell Death Mechanisms in Major Neurodegenerative Diseases. *Int J Mol Sci,* 19(10): 3082.

Chinnaiyan, A.M. (1999). The apoptosome: heart and soul of the cell death machine. *Neoplasia,* 1(1):5-15.

Choi, D.W. (1985). Glutamate neurotoxicity in cortical cell culture is calcium dependent. *Neurosci Lett,* 58: 293–297.

Choi, D.W. (1988). Glutamate Neurotoxicity and Diseases of the Nervous System. *Neuron,* 1: 623-634.

Circu, M.L.,Aw, T.Y. (2010). Reactive oxygen species, cellular redox systems, and apoptosis. *Free Radic Biol Med,* 48(6): 749-762.

Claudio, S. (2003). Unfolding the role of protein misfolding in neurodegenerative diseases. *Nat Rev Neurosci,* 4(1): 49-60.

Cortes, C.J., La Spada, A.R. (2015). Autophagy in polyglutamine disease: Imposing order on disorder or contributing to the chaos? *Mol Cell Neurosci,* 66(Pt A): 5361.

Crighton, D., Wilkinson, S., O'Prey, J., Syed, N., Smith, P., Harrison, P.R., Gasco, M., Garrone, O., Crook, T., Ryan, K.M. (2006).

DRAM, a p53-induced modulator of autophagy, is critical for apoptosis. *Cell,* 126: 121–134.

De Strooper, B., Iwatsubo, T., Wolfe, M.S. (2012). Presenilins and γ-secretase: structure, function, and role in Alzheimer Disease. *Cold Spring Harb Perspect Med*, 2(1): a006304.

Degterev, A., Huang, Z., Boyce, M., Li, Y., Jagtap, P., Mizushima, N., Cuny, G.D., Mitchison, T.J., Moskowitz, M.A., Yuan, J. (2005). Chemical inhibitor of nonapoptotic cell death with therapeutic potential for ischemic brain injury. *Nat Chem Biol*, 1(2): 112-119.

Dexter, D. T. (1989). Basal lipid peroxidation in substantia nigra is increased in Parkinson's disease. *J. Neurochem,* 52: 381-389.

Dong, X.X., Wang, Y., Qin, Z.H. (2009). Molecular mechanisms of excitotoxicity and their relevance to pathogenesis of neurodegenerative diseases. *Acta Pharmacol Sin,* 30(4): 379-387.

Doyle, K.M., Kennedy, D., Gorman, A.M., Gupta, S., Healy, S.J., Samali, A. (2011). Unfolded proteins and endoplasmic reticulum stress in neurodegenerative disorders. *J Cell Mol Med,* 15(10): 2025–2039.

Dringen, R., Hirrlinger, J. (2003). Glutathione pathways in the brain. *Biol Chem.* 384(4): 505-516.

Dringen, R., Pfeiffer, B., Hamprecht, B. (1999). Synthesis of the antioxidant glutathione in neurons: supply by astrocytes of Cys Gly as precursor for neuronal glutathione. *J Neurosci,* 19(2): 562-569.

Du, C., Fang, M., Li, Y., Li, L., Wang, X. (2000). Smac, a mitochondrial protein that promotes cytochrome c-dependent caspase activation by eliminating IAP inhibition. *Cell,* 102(1): 33-42.

Džoljić, E., Grbatinić, I., Kostić, V. (2015). Why is nitric oxide important for our brain? *Funct Neurol,* 30(3): 159-163.

Ebadi, M., Sharma, S.K. (2003). Peroxynitrite and mitochondrial dysfunction in the pathogenesis of Parkinson's disease. *Antioxid Redox Signal,* 5(3): 319-335.

Edinger, A. L., Thompson, C.B. (2004). Death by design: apoptosis, necrosis and autophagy. *Current Opinion in Cell Biology,* 16: 663–669.

Eisenberg-Lerner, K.A. (2012). PKD is a kinase of Vps34 that mediates ROS-induced autophagy downstream of DAPk, *Cell Death Differ,* 19: 788–797.

Elmore, S. (2007). Apoptosis: a review of programmed cell death.*Toxicol Pathol.* 35(4): 495–516.

Enari, M., Sakahira, H., Yokoyama, H., Okawa, K., Iwamatsu, A., Nagata, S. (1998). A caspase-activated DNase that degrades DNA during apoptosis, and its inhibitor ICAD. *Nature,* 391(6662): 43-50.

Estrada Sánchez, A.M., Mejía-Toiber, J., Massieu, L. (2008). Excitotoxic neuronal death and the pathogenesis of Huntington's disease. *Arch Med Res*, 39(3):265-276.

Eyo, U.B., Wu, L.J. (2013). Bidirectional microglia-neuron communication in the healthy brain. *Neural Plast.* 2013: 456857.

Fatokun, A.A., Dawson, V.L., Dawson, T.M. (2014). Parthanatos: mitochondrial-linked mechanisms and therapeutic opportunities. *Br J Pharmacol,* 171(8): 2000–2016.

Festjens, N.,Vanden, Berghe. T.,Cornelis, S.,Vandenabeele, P. (2007). RIP1, a kinase on the crossroads of a cell's decision to live or die. *Cell Death Differ,* 14(3): 400-410.

Fiers, W.,Beyaert, R.,Declercq, W.,Vandenabeele, P. (1999). More than one way to die: apoptosis, necrosis and reactive oxygen damage.*Oncogene,* 18(54): 7719-7730.

Fouillet, A., Levet, C., Virgone, A., Robin, M., Dourlen, P., Rieusset, J., Belaidi, E., Ovize, M., Touret, M., Nataf, S., Mollereau, B. (2012). ER stress inhibits neuronal death by promoting autophagy. *Autophagy.* 8(6): 915-926.

Fridman, J.S.,Lowe, S.W. (2003). Control of apoptosis by p53, *Oncogene,* 22: 9030–9040.

Gage, F.H. (2002). Neurogenesis in the Adult Brain. Fred H. Gage. *Journal of Neuroscience,* 22(3): 612-613.

Gleichmann, M., Mattson, M.P. (2011). Neuronal calcium homeostasis and dysregulation. *Antioxid Redox Signal,* 14(7): 1261–1273.

Greenamyre, J.T., Hastings, T.G. (2004). Parkinson's-divergent causes, convergent mechanisms. *Science,* 304: 1120–1122.

Guan, S.,Xu, J.,Guo, Y.,Ge, D.,Liu, T.,Ma, X.,Cui, Z. (2015). Pyrroloquinoline quinone against glutamate-induced neurotoxicity in cultured neural stem and progenitor cells. *Int J Dev Neurosci,* 42: 37-45.

Guan, Y., Meurer, M., Raghavan, S., Rebane, A., Lindquist, J.R., Santos, S., Kats, I., Davidson, M.W., Mazitschek, R., Hughes, T.E., Drobizhev, M., Knop, M., Shah, J.V. (2015). Live-cell multiphoton fluorescence correlation spectroscopy with an improved large Stokes shift fluorescent protein. *Mol Biol Cell,* 26(11): 2054-2066.

Ha, H.C., Snyder, S.H. (1999). Poly (ADP-ribose) polymerase is a mediator of necrotic cell death by ATP depletion. *Proc Natl Acad Sci USA,* 96:13978-13982.

Halliwell, B. (2009). The wanderings of a free radical. *Free Radic Biol Med,* 46(5): 531-542.

Harman, D. (1992). Role of free radicals in aging and disease. *Ann N Y Acad Sci,* 673: 126-141.

Hill, M.M., Adrain, C., Duriez, P.J., Creagh, E.M., Martin, S.J. (2004). Analysis of the composition, assembly kinetics and activity of native Apaf-1 apoptosomes. *EMBO* J23 (10): 2134-2145.

Hinz, J., Lehnhardt, L., Zakrzewski, S., Zhang, G., Ignatova, Z. (2012). Polyglutamine Expansion Alters the Dynamics and Molecular Architecture of Aggregates in Dentatorubropallidoluysian Atrophy. *J Biol Chem,* 287(3):2068–2078.

Hirsch, T., Marchetti, P., Susin, S.A., Dallaporta, B., Zamzami, N., Marzo, I., Geuskens, M., Kroemer, G. (1997). The apoptosis-necrosis paradox. Apoptogenic proteases activated after mitochondrial permeability transition determine the mode of cell death. *Oncogene,* 15(13): 1573-1581.

Hishikawa, N., Niwa, J., Doyu, M., Ito, T., Ishigaki, S., Hashizume, Y.S. (2003). Dorfin localizes to the ubiquitylated inclusions in Parkinson's disease, dementia with Lewy bodies, multiple system atrophy, and amyotrophic lateral sclerosis. *Am J Pathol*, 163(2): 609- 619.

Hobson, E.V., McDermott, C.J. (2016). Supportive and symptomatic management of amyotrophic lateral sclerosis.*Nature Reviews Neurology*, 12(9): 526-538.

Holler, N.,Zaru, R.,Micheau, O.,Thome, M.,Attinger, A.,Valitutti, S.,Bodmer, J.L.,Schneider, P.,Seed, B.,Tschopp, J. (2000). Fas triggers an alternative, caspase-8-independent cell death pathway using the kinase RIP as effector molecule. *Nat Immunol,*1(6): 489- 495.

Hoozemans, J.J., Veerhuis, R., Van Haastert, E.S., Rozemuller, J.M., Baas, F., Eikelenboom, P. and Scheper, W. (2005). The unfolded protein response is activated in Alzheimer's disease. *Acta Neuropathol*, 110: 165–172.

Hsu, H., Xiong, J., Goeddel, D.V. (1995). The TNF receptor 1- associated protein TRADD signals cell death and NF-kappa B activation. *Cell*, 81(4): 495-504.

Hynd, M.R.,Scott, H.L.,Dodd, P.R. (2004). Glutamate-mediated excitotoxicity and neurodegeneration in Alzheimer's disease. *Neurochem Int*, 45(5): 583-595.

Imai, Y., Soda, M., Inoue, H., Hattori, N., Mizuno, Y., Takahashi, R. (2001). An unfolded putative transmembrane polypeptide, which can lead to endoplasmic reticulum stress, is a substrate of Parkin. *Cell*, 105: 891–902.

Itzhak, Y., Gandia, C., Huang, P.L., Ali, S.F. (1998) Resistance of neuronal nitric oxide synthase-deficient mice to methamphetamine-induced dopaminergic neurotoxicity. *J Pharmacol Exp Ther*, 284(3): 1040-1047.

Jangamreddy, J.R., Los, M.J. (2012). Mitoptosis, a novel mitochondrial death mechanism leading predominantly to activation of autophagy. *Hepat Mon*, 12(8).

Jebelli, J.D., Hooper, C., Garden, G.A., Pocock, J.M. (2011). Emerging roles of p53 in glial cell function in health and disease. *Glia*, 60(4): 515-525.

Jiali, Wang., Thomas, Albers., Christof, Grewer. (2018). Energy Landscape of the Substrate Translocation Equilibrium of Plasma-Membrane Glutamate Transporters. *J. Phys. Chem. B*, 122(1): 28-39.

Johri, A., Beal, M.F. (2012). Mitochondrial dysfunction in neurodegenerative diseases. *J Pharmacol Exp Ther*, 342(3): 619-630.

Joza, N., Susin, S.A., Daugas, E., Stanford, W.L., Cho, S.K., Li, C.Y., Sasaki, T., Elia, A.J., Cheng, H.Y., Ravagnan, L., Ferri, K.F., Zamzami, N., Wakeham, A., Hakem, R., Yoshida, H., Kong, Y.Y., Mak, T.W., Zúñiga-Pflücker, J.C., Kroemer, G., Penninger, J.M. (2001). Essential role of the mitochondrial apoptosis-inducing factor in programmed cell death. *Nature*, 410(6828): 549-554.

Kametani, F., Hasegawa, M. (2018). Reconsideration of Amyloid Hypothesis and Tau Hypothesis in Alzheimer's Disease. *Front Neurosci*, 12: 25.

Kandimalla, R., Reddy, P.H. (2017). Therapeutics of Neurotransmitters in Alzheimer's Disease. *J Alzheimers Dis*, 57(4):1049-1069.

Kell, D.B. (2010). Towards a unifying, systems biology understanding of large-scale cellular death and destruction caused by poorly liganded iron: Parkinson's, Huntington's, Alzheimer's, prions, bactericides, chemical toxicology and others as examples. *Archives of Toxicology*, 84(11): 825–889.

Kerr, J.F. (2002). History of the events leading to the formulation of the apoptosis concept. *Toxicology*, 182: 471-474.

Kerr, J.F., Wyllie, A.H., Currie, A.R. (1972). Apoptosis: a basic biological phenomenon with wide-ranging implications in tissue kinetics. *Br J Cancer,* 26(4): 239-257.

Kischkel, F.C., Hellbardt, S., Behrmann, I., Germer, M., Pawlita, M., Krammer, P.H., Peter, M.E. (1995). Cytotoxicity-dependent APO-1 (Fas/CD95)-associated proteins form a death-inducing signaling complex (DISC) with the receptor. *EMBO J,* 14(22): 5579-5588.

Kobayashi, S. (2016). Choose delicately and Reuse Adequately: The Newly Revealed Process of Autophagy. *Biol Pharm Bull,* 38(8): 1098-1103.

Korhonen, L., Lindholm, D. (2004). The ubiquitin proteasome system in synaptic and axonal degeneration: a new twist to an old cycle. *J Cell Biol,* 165: 27–30.

Kouroku, Y., Fujita, E., Tanida, I., Ueno, T., Isoai, A., Kumagai H., Ogawa, S., Kaufman, R.J., Kominami, E., Momoi, T. (2007). ER stress (PERK/eIF2alpha phosphorylation) mediates the polyglutamine-induced LC3 conversion, an essential step for autophagy formation. *Cell Death Differ,* 14(2): 230-239.

Kovacs, D.M., Mancini, R., Henderson, J., Na, S.J., Schmidt, S.D., Kim, T.W., Tanzi, R.E. (1999). Staurosporine-induced activation of caspase-3 is potentiated by presenilin 1 familial Alzheimer's disease mutations in human neuroglioma cells. *J Neurochem,* 73(6):2278-2285.

Lang, A.E., Lozano, A.M. (1998). Parkinson's disease. First of two parts. *N Engl J Med,* 339(15): 1044-1053.

Lau, A., Michael Tymianski, M. (2010). Glutamate receptors, neurotoxicity and neurodegeneration. *Pflugers Arch - Eur J Physiol,* 460: 525-542.

Lee, J.S., Li, Q., Lee, J.Y., Lee, S.H., Jeong, J.H., Lee, H.R., Chang, H., Zhou, F.C., Gao, S.J., Liang, C., Jung, J.U. (2009). FLIP-mediated autophagy regulation in cell death control. *Nat. Cell Biol,* 11: 1355–1362.

Lee, K.M., MacLean, A.G. (2015). New advances on glial activation in health and disease.*World J Virol*, 4(2): 42-55.

Levine, B., Kroemer, G. (2008). Autophagy in the pathogenesis of disease, *Cell*, 132: 27–42.

Levine, S.M. (1997). Iron deposits in multiple sclerosis and Alzheimer's disease brains. *Brain Research*, 760(1-2): 298–303.

Lewen, A., Matz, P. & Chan, P. H. (2000). Free radical pathways in CNS injury. *J. Neurotrauma*, 17: 871–890.

Lewerenz, J., Maher, P. (2015). Chronic Glutamate Toxicity in Neurodegenerative Diseases-What is the Evidence?*Front Neurosci*, 9: 469.

Li, C.J., Friedman, D.J., Wang, C., Metelev, V., Pardee, A.B. (1995). Induction of apoptosis in uninfected lymphocytes by HIV-1 Tat protein. *Science*, 268(5209): 429-431.

Lindholm, D., Wootz, H. Korhonen, L. (2006). ER stress and neurodegenerative diseases. *Cell Death and Differentiation*, 13: 385–392.

Lipinski, M.M., Yuan, J. (2004) Mechanisms of cell death in polyglutamine expansion diseases. *Curr. Opin. Pharmacol*, 4: 85–90.

Magrinelli, F., Picelli, A., Tocco, P., Federico, A., Roncari, L., Smania, N., Zanette, G., Tamburin, S. (2016). Pathophysiology of Motor Dysfunction in Parkinson's Disease as the Rationale for Drug Treatment and Rehabilitation. *Parkinsons Dis*, 2016: 9832839.

Malhotra, J.D., Kaufman, R.J. (2007). The endoplasmic reticulum and the unfolded protein response. *Semin Cell Dev Biol*, 18(6): 716-731.

Martinvalet, D., Zhu, P., Lieberman, J. (2005). Granzyme A induces caspase-independent mitochondrial damage, a required first step for apoptosis. *Immunity*, 22(3): 355-370.

Matilla-Dueñas, A., Ashizawa, T., Brice, A., Magri, S., McFarland, K.N., Pandolfo, M., Pulst, S.M., Riess, O., Rubinsztein, D.C., Schmidt, J., Schmidt, T., Scoles, D.R., Stevanin, G., Taroni, F., Underwood, B.R., Sánchez, I. (2014). Consensus paper:

pathological mechanisms underlying neurodegeneration in spinocerebellar ataxias. *Cerebellum,* 13(2): 269-302.

Melton, L.M., Keith, A.B., Davis, S., Oakley, A.E., Edwardson, J.A., Morris, C.M. (2003). Chronic glial activation, neurodegeneration, and APP immunoreactive deposits following acute administration of double-stranded RNA. *Glia,* 44(1): 1-12.

Migaud, M.,Batailler, M.,Segura, S.,Franceschini, D.I.,Pillon, D. (2010). Emerging new sites for adult neurogenesis in the mammalian brain: a comparative study between the hypothalamus and the classical neurogenic zones. *Eur. J. Neurosci,* 32:2042-2052.

Mizushima, N., Komatsu, M. (2011). Autophagy:renovation of cells and tissues. *Cell,* 147(4): 728-741.

Mizushima, N., Levine, B. (2010). Autophagy in mammalian development and differentiation,*Nat. Cell Biol,* 12: 823–830.

Mizushima, N., Levine, B., Cuervo, A.M., Klionsky, D.J. (2008). Autophagy fights disease through cellular self-digestion, *Nature,* 451: 1069–1075.

Mohmmad, Abdul. H., Sultana, R., Keller, J.N., St Clair, D.K., Markesbery, W.R. (2006). Butterfield DA. Mutations in amyloid precursor protein and presenilin-1 genes increase the basal oxidative stress in murine neuronal cells and lead to increased sensitivity to oxidative stress mediated by amyloid beta-peptide (1-42), HO and kainic acid: implications for Alzheimer's disease. *J Neurochem,* 96(5):1322-35.

Morishima, N., Nakanishi, K., Tsuchiya, K., Shibata, T. Seiwa, E. (2004). Translocation of Bim to the endoplasmic reticulum (ER) mediates ER stress signaling for activation of caspase-12 during ER stress-induced apoptosis. *J. Biol. Chem,* 279: 50375–50381.

Morrison, R.S.,Kinoshita, Y. (2000). The role of p53 in neuronal cell death. *Cell Death Differ,* 7(10): 868-879.

Mrakovcic, M., Fröhlich, L.F. (2018). p53-Mediated Molecular Control of Autophagy in Tumor Cells. *Biomolecules,* 8(2): 14.

Nakagawa, T., Yuan, J. (2000). Cross-talk between two cysteine protease families. Activation of caspase-12 by calpain in apoptosis. *J. Cell Biol,* 150: 887–894.

Nelson, N. (1999). Metal ion transporters and homeostasis. *The EMBO Journal,* 18(16): 4361–4371.

Nemes, Z. Jr., Friis, R.R., Aeschlimann, D., Saurer, S., Paulsson, M., Fésüs, L. (1996). Expression and activation of tissue transglutaminase in apoptotic cells of involuting rodent mammary tissue. *Eur J Cell Biol,* 70(2): 125-133.

Nikoletopoulou, V., Markaki, M., Palikaras, K., Tavernarakis, N. (2013). Crosstalk between apoptosis, necrosis and autophagy. *Biochimica et Biophysica Acta,* 16987.

Nisbet, R.M., Polanco, J.C., Ittner, L.M., Götz, J. (2014). Tau aggregation and its interplay with amyloid-β.*Acta Neuropathol,* 129(2): 207–220.

Nishitoh, H., Matsuzawa, A., Tobiume, K., Saegusa, K., Takeda, K., Inoue, K., Hori, S., Kakizuka, A., Ichijo, H. (2002). ASK1 is essential for endoplasmic reticulum stress-induced neuronal cell death triggered by expanded polyglutamine repeats. *Genes Dev,* 16: 1345–1355.

Ozawa, T. (1995). Mechanism of somatic mitochondrial DNA mutations associated with age and diseases. *Biochim Biophys Acta,* 1271(1): 177-189.

Paschen, W., Frandsen, A. (2001). Endoplasmic reticulum dysfunction a common denominator for cell injury in acute and degenerative diseases of the brain? *J. Neurochem,* 79: 719–725.

Pasinelli, P., Borchelt, D.R., Houseweart, M.K., Cleveland, D.W., Brown, R.H. Jr. (1998). Caspase-1 is activated in neural cells and tissue with amyotrophic lateral sclerosis-associated mutations in copper-zinc superoxide dismutase [published correction appears in *Proc Natl Acad Sci U S A,* 95(26): 15763–15768.

Pattingre, S., Tassa, A., Qu, X., Garuti, R., Liang, X.H., Mizushima, N., Packer, M., Schneider, M.D., Levine, B. (2005). Bcl-2 antiapoptotic proteins inhibit Beclin 1-dependent autophagy, *Cell,* 122: 927–939.

Paul, V., Ekambaram, P. (2011). Involvement of nitric oxide in learning& memory processes. *Indian J Med Res,* 133: 471-478.

Paweletz, N. (2001). Walther Flemming: pioneer of mitosis research. *Nat Rev Mol Cell Biol,* 2(1):72-75.

Pósfai, B.,Cserép, C.,Orsolits, B.,Dénes, Á. (2019). New Insights into Microglia-Neuron Interactions: A Neuron's Perspective. *Neuroscience,* 405: 103-117.

Radi, R. (2018). *Oxygen radicals, nitric oxide, and peroxynitrite: Redox pathways in molecular medicine.* 115 (23): 5839-5848.

Raevskiĭ, K.S.,Bashkatova, V.G.,Vanin, A.F. (2000). The role ofnitric oxideinbrainglutaminergic pathology. *Vestn Ross Akad Med Nauk,*(4): 11-15.

Ranganathan, S., Bowser, R. (2010). p53 and Cell Cycle Proteins Participate in Spinal Motor Neuron Cell Death in ALS. *Open Pathol J,*4: 11-22.

Rao, R.V., Ellerby, H.M. Bredesen, D.E. (2004) Coupling endoplasmic reticulum stress to the cell death program. *Cell Death Differentiation,* 11: 372–380.

Reddy, P.H., Shirendeb, U.P. (2011). Mutant huntingtin, abnormal mitochondrial dynamics, defective axonal transport of mitochondria, and selective synaptic degeneration in Huntington's disease. *Biochim Biophys Acta,* 1822(2).

Reeve, A., Simcox, E., Turnbull, D. (2014). Ageing and Parkinson's disease: why is advancing age the biggest risk factor? *Ageing Res Rev,* 14: 19-30.

Reitz, C., Brayne, C., Mayeux, R. (2011). Epidemiology of Alzheimer disease. *Nat Rev Neurol,* 7(3):137-52.

Ricci, G., Volpi, L., Pasquali, L., Petrozzi, L., Siciliano, G. (2009). Astrocyte-neuron interactions in neurological disorders. *J Biol Phys,* 35(4): 317-336.

Ricci, G., Volpi, L., Pasquali, L., Petrozzi, L., Siciliano, G. (2009). Astrocyte-neuron interactionsin neurological disorders. *J Biol Phys*, 35(4): 317-336.

Ross, C.A., Poirier, M.A. (2004). Protein aggregation and neurodegenerative disease. *Nat Med,* 10 Suppl: S10-7.

Rossi, D., Volterra, A. (2009). Astrocytic dysfunction: insights on therolein neurodegeneration. *Brain Res Bull*, 80(4-5): 224-232.

Rowland, L.P., Shneider, M.D. (2001) Amyotrophic lateral sclerosis. *N Engl J Med,* 344: 1688–1700.

Ruiz, A., Matute, C., Alberdi, E. (2009). Endoplasmic reticulum Ca2+ release through ryanodine and IP3 receptors contributes to neuronal excitotoxicity. *Cell Calcium,* 46(4): 273–281.

Zorov, D.B., Juhaszova, M., Sollott, S.J. (2014). Mitochondrial reactive oxygen species (ROS) and ROS-induced ROS release. *Physiol Rev,* 94(3): 909-950.

Saelens, X., Festjens, N., Vande Walle, L., van Gurp, M., van Loo, G., Vandenabeele, P. Toxic proteins released from mitochondria in cell death. *Oncogene,* 23(16): 2861-2874.

Salvador, G.A., Uranga, R.M., Giusto, N.M. (2010). Iron and mechanisms of neurotoxicity. *Int J Alzheimers Dis,* 2011: 720658.

Schuler, M., Green, D.R. (2001). Mechanisms of p53-dependent apoptosis. *Biochem Soc Trans*, 29: 684–688.

Seitz, G., Stegmann, H.B., Jager, H.H., Schlude, H.M., Wolburg, H., Roginsky, V.A., Niethammer, D., Bruchelt, G. (2000). Neuroblastoma cells expressing the noradrenaline transporter are destroyed more selectively by 6-fluorodopamine than by 6-hydroxydopamine. *J. Neurochem*, 75: 511–520.

Selkoe, D.J. (2001). Alzheimer's disease: genes, proteins, and therapy. *Physiol. Rev,* 81: 741–766.

Shah, S., Lubeck, E., Schwarzkopf, M., He, T.F., Greenbaum, A., Sohn, C.H., Lignell, A., Choi, H.M., Gradinaru, V., Pierce, N.A., Cai, L. (2006). Single molecule RNA detection at depth by hybridization chain reaction and tissue hydrogel embedding and clearing. *Development* 143: 2862–2867.

Shah, S.A.,Amin, F.U.,Khan, M.,Abid, M.N.,Rehman, S.U.,Kim, T.H.,Kim, M.W.,Kim, M.O. (2016). Anthocyanins abrogateglutamate-inducedAMPK activation, oxidative stress, neuroinflammation, and neurodegeneration in postnatal rat brain.*J Neuroinflammation,* 13(1): 286.

Shaw, P.J.,Ince, P.G. (1997). Glutamate, excitotoxicity and amyotrophic lateral sclerosis. *J Neurol,* 244 Suppl 2: S3-14.

Shirendeb, U.P., Calkins, M.J., Manczak, M. (2011). Mutant huntingtin's interaction with mitochondrial protein Drp1 impairs mitochondrial biogenesis and causes defective axonal transport and synaptic degeneration in Huntington's disease. *Hum Mol Genet,* 21(2): 406–420.

Simić, G., Babić Leko, M., Wray, S. (2016). Tau Protein Hyperphosphorylation and Aggregation in Alzheimer's Disease and Other Tauopathies, and Possible Neuroprotective Strategies. *Biomolecules,* 6(1): 6.

Singh, S. (2017). Environmental Toxins and Neurodegeneration, Chapter 17, *Environmental Sci. & Eng,* Vol. 6: Toxicology, Studium Press LLC.

Singh, S.,Swarnkar, S.,Goswami, P.,Nath, C. (2011). Astrocytes and microglia: responses to neuropathological conditions. *Int J Neurosci,* 121(11): 589-597.

Slee, E.A., Adrain, C., Martin, S.J. (2001). Executioner caspase-3, -6, and -7 perform distinct, non-redundant roles during the demolition phase of apoptosis. *J Biol Chem,* 276(10): 7320-7326.

Soto, C. (2003). Unfolding the role of protein misfolding in neurodegenerative diseases. *Nat. Rev. Neurosci,* 4: 49–60.

Spalding, K.L.,Bergmann, O.,Alkass, K.,Bernard, S.,Salehpour, M.,Huttner, H.B.,Boström, E.,Westerlund, I.,Vial, C.,Buchholz, B.A.,Possnert, G.,Mash, D.C.,Druid, H.,Frisén, J. (2013). Dynamics of hippocampal neurogenesis in adult humans. *Cell,*153(6): 1219-1227.

Stefanis, L. (2012). α-Synuclein in Parkinson's disease. *Cold Spring Harb Perspect Med,* 2(2): a009399.

Tang, T.S., Tu, H., Chan, E.Y., Maximov, A., Wang, Z., Wellington, C.L., Hayden, M.R., Bezprozvanny, I. (2003) Huntingtin and huntingtin-associated protein 1 influence neuronal calcium signaling mediated by inositol-(1,4,5) triphosphate receptor type 1. *Neuron,* 39: 227–239.

Teresa R, Taylor-Whiteley., Christine L, Le Maitre., James A. Duce, Caroline F. Dalton, David P. Smith. (2019). Recapitulating Parkinson's disease pathology in a three-dimensional human neural cell culture model. *Disease Models & Mechanisms,* 12: dmm038042.

Turner, B.J., Atkin, J.D., Farg, M.A., Zang, D.W., Rembach, A., Lopes, E.C., Patch, J.D., Hill, A.F., and Cheema, S.S. (2005). Impaired Extracellular Secretion of Mutant Superoxide Dismutase 1 Associates with Neurotoxicity in Familial Amyotrophic Lateral Sclerosis.*J. Neurosci,* 25(1): 108 –117.

Uttara, B., Singh, A.V., Zamboni, P., Mahajan, R.T. (2009). Oxidative stress and neurodegenerative diseases: a review of upstream and downstream antioxidant therapeutic options. *Curr Neuropharmacol,* 7(1): 65–74.

Vanlangenakker, N.,Vanden, Berghe. T.,Krysko, D.V.,Festjens, N.,Vandenabeele, P. (2008). Molecular mechanisms and pathophysiology of necrotic cell death. *Curr Mol Med,* 8(3): 207-220.

Wajant, H. (2002). The Fas signaling pathway: more than a paradigm. *Science,* 296(5573):1635-1636.

Wootz, H., Hansson, I., Korhonen, L., Napankangas, U., Lindholm, D. (2004). Caspase-12 cleavage and increased oxidative stress during motoneuron degeneration in transgenic mouse model of ALS. *Biochem. Biophys. Res. Commun,* 322: 281–286.

Worth, A., Thrasher, A.J., Gaspar, H.B. (2006). Autoimmune lymphoproliferative syndrome: molecular basis of disease and clinical phenotype. *Br J Haematol,* 133(2):124-140.

Xie, Z., D.J. Klionsky, D.J. (2007). Autophagosome formation: core machinery and adaptations, *Nat. Cell Biol,* 9: 1102–1109.

Yeo, W., Gautier, J. (2004). Early neural cell death: dying to become neurons. *Developmental Biology,* 274(2): 233-244.

Yuste, J.E.,Tarragon, E.,Campuzano, C.M.,Ros-Bernal, F. (2015). Implications of glial nitric oxide in neurodegenerative diseases. *Front Cell Neurosci,* 9: 322.

Zecca, L.,Youdim, M.B.,Riederer, P.,Connor, J.R.,Crichton, R.R. (2004). Iron, brain ageing and neurodegenerative disorders. *Nat Rev Neurosci,* 5(11):863-873.

Zeiss, C.J. (2003). The apoptosis-necrosis continuum: insights from genetically altered mice. *Vet Pathol,* 40(5):481-495.

Zeng, X., Overmeyer, J.H., Maltese, W.A. (2006). Functional specificity of the mammalian Beclin-Vps34 PI 3-kinase complex in macroautophagy versus endocytosis and lysosomal enzyme trafficking, *J. Cell Sci,* 119: 259-270.

Zhang, Y.W., Thompson, R., Zhang, H., Xu, H. (2011). APP processing in Alzheimer's disease. *Mol Brain,* 4: 3.

Zhou, X., Hollern, D., Liao, J., E Andrechek, E.,Wang, H. (2013). NMDA receptor-mediated excitotoxicity depends on the coactivation of synaptic and extrasynaptic receptors. *Cell Death & Disease,* 4: e560.

Zucca, F.A., Segura-Aguilar, J., Ferrari, E. (2015). Interactions of iron, dopamine and neuromelanin pathways in brain aging and Parkinson's disease. *Prog Neurobiol,* 155: 96–119.

ABOUT THE EDITOR

Gokul Krishna, PhD
Postdoctoral Research Associate,
University of Arizona Phoenix, US

Dr. Gokul Krishna is a Postdoctoral Research Associate in the Department of Child Health at the University of Arizona Phoenix. He graduated from Rajiv Gandhi University of Health Sciences in 2007 with a Bachelor of Pharmacy in Pharmaceutical Sciences and Master of Pharmacy in Pharmacology in 2010 from the same institution. He earned his Ph.D. in Biochemistry from University of Mysore in 2016 and continued as a postdoctoral fellow in the Neurotrophic Research Laboratory at the University of California Los Angeles. His research goal is to explore how traumatic brain injury cause circuit dysfunction that may influence the development of emotional and behavioral deficits. Previous projects also explored - 1) neuronal adaptation to brain trauma by focusing on the role of neurotrophic factors as major regulators of neuronal plasticity and; 2) oxidative aspects of neurodegenerative diseases, in particular, the role of environmental agents (e.g. pesticides) to influence the onset of neurodegenerative diseases and interventions to mitigate the neurotoxic responses.

INDEX

#

3-nitrotyrosine, 22

A

abuse, 23, 50
acetylcholine, 12, 24, 31, 32, 51, 86
acetylcholinesterase, 11, 30
acid, 8, 39, 55, 60, 80, 120
acrylamide, 22, 47, 48, 49, 50
action potential, 8
acute, 24, 52, 111, 120, 121
AD, ix, 54, 56, 57, 58, 59, 66, 72, 74, 85, 86, 87, 95, 102, 103, 104
adaptive immune response, 98
additives, 10, 14
adenine, 11, 12, 91
adhesion, 99
adipose tissue, 21
adulthood, 12, 14, 27, 29, 32, 35
adverse, vii, 1, 3, 6, 7, 8, 13, 18, 22, 23, 29, 34, 93, 95, 96
adverse effects, 3, 6, 8, 13, 18, 23, 34, 96

age, iv, v, 4, 7, 24, 45, 52, 85, 86, 87, 90, 98, 104, 106, 110, 121, 122
aggregation, viii, 12, 33, 34, 54, 56, 58, 79, 80, 83, 86, 87, 96, 101, 110, 121, 123, 124
alpha-synuclein, 55, 79, 80, 81, 82, 83, 84, 87, 102, 105
alters, 13, 23, 30, 33, 37, 38, 39, 41, 52
amino, 13, 17, 23, 33
amino acid, 13, 17, 23, 33
amyloid beta, 55, 80, 120
amyloid beta peptide, 55
amyloid fibril formation, 84
amyotrophic lateral sclerosis, 110, 116, 121, 124
antibody, 11, 59, 62, 63
antioxidant, 14, 17, 19, 43, 113, 125
antisocial personality, 16
anxiety -like, 13
apoptosis, 21, 29, 39, 49, 85, 88, 89, 91, 93, 99, 102, 110, 111, 112, 113, 114, 115, 117, 119, 120, 121, 123, 124, 126
apoptotic mechanisms, 99
arsenic, 18, 40, 41, 42, 46
aspartate, 3, 17, 21, 24, 51, 94
astrocyte, 10, 58, 122, 123

Index

astrocytes, 21, 31, 98, 113
atrophy, 56, 107, 116
autism, 2, 3, 7, 11, 25, 26, 27, 28, 30, 31
autoimmune disease, 90
axonal degeneration, 118

B

bactericides, 117
behavioral assessment, 36
behaviors, 14, 36, 41, 45, 50
beneficial effect, 20, 98
biological fluids, 83
biological processes, 16
biotransformation, 10
birth, 11, 16, 23, 30, 49
blood, 6, 13, 16, 21, 26, 37, 58, 97
blood -brain barrier, 26
blood-brain barrier, 6, 13, 26
brain, vii, 1, 2, 4, 5, 6, 9, 10, 11, 13, 14, 15, 17, 19, 21, 23, 26, 28, 29, 30, 31, 32, 33, 34, 35, 36, 38, 39, 41, 43, 44, 45, 46, 49, 50, 55, 58, 74, 78, 81, 82, 84, 85, 86, 95, 97, 98, 99, 102, 104, 105, 106, 109, 110, 111, 113, 114, 119, 120, 121, 123, 124, 126, 127
brain -derived neurotrophic factor, 2, 13
brainstem, 12, 38, 50

C

Ca^{2+}, 14, 19, 21, 90, 123
calcium, 17, 37, 93, 94, 95, 99, 101, 102, 104, 107, 111, 112, 115, 125
cAMP response element-binding protein, 13
cancer, 18, 40, 90, 118
carbohydrate, 22, 99
carboxylic acid, 8
carcinogenicity, 48
Caspase-8, 89
caspases, 89, 107
catalysis, 89
cell culture, 34, 61, 83, 112, 125
cell cycle, 10, 12, 14, 29
cell death, ix, 12, 14, 19, 56, 88, 89, 92, 98, 101, 108, 109, 112, 113, 114, 115, 116, 117, 118, 119, 120, 121, 122, 123, 125, 126
cell differentiation, 14, 65
cellular energy, 17, 24, 86, 97, 100, 106, 107
central nervous system, 3, 48, 50, 51, 52, 57, 85, 111
central nervous system (CNS), 3, 57
cerebellar development, 44
cerebellum, 11, 20, 23, 27, 49, 120
chemical, vii, viii, 1, 3, 6, 12, 14, 24, 57, 58, 73, 86, 117
chemical properties, 58
chemicals, vii, 1, 2, 5, 6, 7, 14, 25, 26, 35
children, 3, 6, 14, 16, 21, 24, 25, 26, 30, 35, 37, 45, 46
cholesterol, 54, 56, 57, 79
cholinergic, 10, 15, 24, 29, 35, 86, 104
cholinesterase, 10
circuits, 13, 25
cleavage, 12, 93, 110, 125
CNS, 3, 17, 29, 44, 57, 94, 97, 98, 119
cognition, 1, 18, 24, 33, 96
cognitive, vii, 3, 11, 14, 15, 16, 21, 22, 31, 33, 37, 45, 46, 56, 86, 104
cognitive deficit, 16, 21, 22
cognitive impairment, 11, 31, 46
cognitive performance, 15
communication, 57, 114
composition, 54, 57, 73, 79, 115
compounds, vii, 8, 13, 14, 106
copper, 45, 63, 96, 99, 121
cortex, 12, 13, 17, 18, 33, 39, 40, 56, 79, 81, 106
corticosterone, 18, 42
critical period, 3
culture, 13, 61, 64, 65, 67, 68, 73

culture media, 61, 64, 67, 68, 73
cytoarchitecture, 17
cytochrome, 10, 12, 15, 28, 89, 113
cytochrome C, 12
cytoplasm, 75, 91, 100, 103

D

decoding, 102
defects, 11, 28, 31, 90
degradation, 9, 14, 55, 88, 101, 107
dementia, 56, 81, 116
dendritic spines, 15, 35
depolarization, 8, 100
deposits, 119, 120
depression, 41
destruction, 89, 117
detoxifying, 19
developing brain, 5, 11, 26
development, vii, ix, 3, 4, 5, 7, 10, 11, 12, 13, 14, 17, 18, 20, 23, 25, 26, 27, 29, 32, 33, 34, 36, 38, 41, 44, 49, 50, 54, 81, 88, 89, 120, 127
developmental, v, viii, 1, 2, 3, 4, 5, 6, 7, 8, 9, 10, 11, 14, 15, 17, 18, 19, 20, 24, 25, 26, 28, 29, 30, 31, 32, 33, 34, 35, 36, 37, 38, 39, 40, 41, 42, 44, 45, 46, 47, 50, 90, 126
Diazinon, 11, 32
diet, viii, 1, 16, 32, 49
dietary intake, 22
digestion, 88, 92, 120
diseases, v, vii, viii, ix, 2, 3, 4, 7, 26, 54, 55, 56, 58, 72, 75, 81, 83, 85, 86, 87, 90, 92, 96, 98, 99, 101, 103, 104, 108, 109, 110, 112, 113, 117, 119, 121, 124, 125, 126, 127
distribution, 71, 81, 108
DNA, 5, 9, 17, 22, 38, 89, 91, 99, 102, 110, 112, 114
DNA damage, 89, 91, 102, 112

DNA repair, 91
DNase, 114
dopamine, 2, 8, 9, 12, 22, 32, 33, 58, 65, 80, 82, 83, 86, 96, 102, 105, 111, 126
dopaminergic, 7, 8, 10, 13, 23, 27, 28, 29, 34, 56, 58, 60, 65, 86, 87, 105, 116
double bonds, 106
drinking water, 16, 17, 18, 19, 20, 45
dysfunction, ix, 7, 20, 26, 55, 66, 100, 101, 106, 119, 121, 123, 127

E

early-life, viii, 2, 4, 16, 26
elderly population, ix
electron, 21, 60, 63, 71, 96, 99
electron microscopy, 60, 71
electrophoretic separation, 22
elongation, 93
embryonic stem cells, 38
endothelial cells, 97
energy metabolism, 17, 24, 97, 100, 106, 107
environment, vii, viii, 1, 5, 6, 14, 18, 19, 26, 74, 92
environmental factors, 3
environmental impact, 4
enzymes, 10, 14, 15, 17, 19, 39, 74, 95, 99
epidemiologic, vii, 4, 7
epidemiologic studies, 7
epigenetic modification, 5
equilibrium, 94, 98
ER Stress, 100
evidence, ix, 16, 17, 19, 24, 26, 54, 56, 59
excitotoxicity, 19, 59, 94, 107, 109, 113, 116, 123, 124, 126
exclusion, 60, 78
exosomes, viii, 54, 55, 57, 58, 69, 70, 71, 72, 73, 74, 75, 76, 77, 79, 80, 82, 83, 84
experimental condition, 72
exploration, 11, 23

exposure, vii, 1, 6, 7, 8, 10, 11, 12, 15, 16, 17, 18, 19, 20, 21, 22, 23, 24, 27, 28, 29, 30, 31, 32, 33, 34, 35, 36, 37, 38, 39, 40, 41, 42, 43, 44, 45, 46, 47, 49, 51, 52, 64, 67, 100, 106
extraction, 60, 61, 63, 67, 72
eye, 19

F

families, 121
ferritin, 97, 109
fetal, 5, 14, 19, 23, 29, 33, 48, 49, 50, 60
fetal abnormalities, 19
fibrillation, 71, 80, 81
fibroblast growth factor, 16, 29
fission, 100, 109
fluorescence, 66, 67, 115
formaldehyde, 62, 63
formation, ix, 5, 13, 22, 71, 72, 74, 84, 91, 92, 101, 104, 118, 126
frontal cortex, 40, 56, 79, 81
functional changes, 111
fungicide, 13, 44

G

gastrointestinal tract, 97
gene expression, 14, 18, 19, 30, 40, 42, 52, 107, 110
genes, 10, 13, 15, 16, 18, 31, 42, 52, 83, 120, 123
genetic, vii, 1, 3, 4, 7, 24, 75
genetic factors, 3, 24
genetic predisposition, 3
gestation, 2, 4, 17, 20, 21, 35, 37, 49
gestation day, 2, 4, 21
glia, 15, 22, 47, 98, 117, 120
glial cells, 96, 97
gliosis, 12, 21
glucocorticoid, 18, 42

glucocorticoid receptor, 18, 42
glutamate, 15, 16, 19, 21, 24, 36, 51, 81, 87, 94, 95, 99, 107, 110, 112, 115, 116, 117, 118, 119, 124
glutamatergic, 13, 43, 111
glutamic acid, 17
glutathione, 14, 19, 43, 113
growth, 9, 15, 23, 35

H

harmful effects, 2, 24, 59
hazardous substance, 18
HD, 43, 61, 85, 87, 95, 96, 107
healing, 90
health, 3, 6, 22, 26, 35, 117, 119
hippocampus, 18, 20, 38, 39, 40, 41, 44, 47, 51, 52, 56, 88
histone, 5, 9, 16, 40
homeostasis, 6, 24, 44, 81, 95, 97, 101, 105, 112, 115, 121
homovanillic acid, 9
human, viii, 1, 14, 15, 20, 22, 31, 35, 38, 49, 50, 58, 64, 65, 79, 80, 86, 102, 105, 118, 125
human body, 86
humans, vii, 4, 6, 18, 22, 23, 25, 32, 40, 87, 88, 92, 125
hybridization, 123
hydrogen peroxide, 97, 99
hyperactivity, 3, 14, 35, 40
hypothalamus, 21, 88, 120
hypoxia, 102, 107

I

idiopathic, 11, 28, 31, 97
images, 61, 63, 64, 67, 69, 70, 76, 77
immune function, 14, 31
immune response, 24, 99
immunofluorescence, 62

immunoreactive, 11, 24, 120
immunoreactivity, 8, 65, 66, 72
immunosuppression, 11
in utero, 4, 20, 28, 40, 46
in vitro, 10, 12, 13, 29, 34, 42, 59
incidence, 3, 7
induction, 19, 36, 65
industrial chemicals, 25
inflammation, 59, 109
ingestion, 18, 21, 87
inhibition, 10, 17, 19, 24, 36, 91, 92, 96, 103, 113
inhibitor, 92, 93, 95, 113, 114
injury, 43, 99, 100, 105, 113, 119, 121
innate immunity, 98
insecticide, 13, 28, 30, 31
insecticides, 8, 9, 27, 28, 29, 31, 87
insulin resistance, 84
integrity, 71, 73
intracellular calcium, 100, 107
ion channels, 8, 27, 94
ion transport, 121
iron, 46, 87, 94, 96, 99, 110, 117, 119, 123, 126

J

juvenile delinquency, 16, 37
juveniles, 22

K

kidney, 93
kinetics, 115, 118

L

laboratory studies, 4
lactation, 9, 37
later life, 3, 7, 26

lateral sclerosis, 85, 106, 123
lead, ix, 4, 6, 16, 17, 35, 37, 38, 39, 40, 85, 88, 89, 91, 93, 96, 97, 98, 100, 104, 108, 116, 120
learning, 3, 13, 17, 33, 46, 122
lesions, 35
Lewy bodies, 7, 56, 81, 116
life expectancy, ix
lifetime, viii
lipid metabolism, 15
lipid peroxidation, 20, 113
lipid rafts, viii, 54, 55, 56, 57, 58, 65, 67, 68, 71, 72, 73, 74, 75, 78, 79, 80, 81
locomotor, 9, 13, 20
locus, 105
long-term, 11, 21, 23, 32, 45
lung cancer, 40
lymphoid, 90

M

machinery, 103, 126
macromolecules, 99
magnetic resonance, 45, 51
magnetic resonance imaging, 45, 51
major depression, 42
manganese, 21, 44, 45, 46, 47
Manganese, 20, 21, 44, 45, 47
manufacturing, 14, 16
maternal, 14, 23, 30, 35, 43, 44, 47, 49
maturation, 2, 3, 10, 17, 23
mechanisms, vii, viii, ix, 2, 5, 6, 7, 17, 19, 27, 29, 34, 41, 42, 54, 55, 56, 57, 59, 81, 85, 87, 93, 94, 99, 104, 105, 108, 110, 112, 113, 114, 115, 119, 120, 123, 125
membranes, viii, 62, 73, 80, 83, 84
memory, 1, 3, 13, 17, 18, 24, 33, 41, 46, 96, 122
memory, cognition, 1, 3, 13, 17, 18, 24, 33, 41, 46, 96, 122
menopause, 74, 78

mental, 19, 87
metabolism, 6, 10, 15, 17, 24, 26, 36, 39, 41, 96, 99, 104, 106, 107
metabolites, 10, 15, 28, 36, 106
metal ions, 97, 99, 100
metals, vii, 1, 6, 16, 45, 46, 97, 99, 112
methamphetamine, 106, 116
methylation, 5, 9, 38, 40
methylmercury, 15, 19, 42, 43, 44
mice, 11, 15, 19, 21, 24, 28, 29, 30, 33, 36, 37, 39, 42, 43, 44, 47, 51, 52, 96, 105, 107, 116, 126
microglia, 10, 21, 34, 81, 82, 98, 114, 122, 124
microscope, 63
microscopy, 61, 63, 64, 67, 69, 71
microstructures, 54, 56, 58, 71, 74
mitochondria, 13, 43, 46, 89, 93, 95, 97, 106, 107, 108, 122, 123
mitochondrial, ix, 2, 7, 9, 10, 17, 20, 21, 24, 49, 59, 86, 89, 93, 96, 99, 100, 102, 104, 106, 107, 108, 109, 110, 111, 112, 113, 114, 115, 117, 119, 121, 122, 123, 124
mitochondrial dysfunction, ix, 2, 7, 10, 20, 49, 99, 100, 102, 104, 106, 107, 113, 117
mitogen, 17, 19, 101
mitogen-activated protein kinase, 17, 101
models, 2, 4, 11, 16, 19, 22, 25, 26, 27, 34, 82, 93, 106, 125
modifications, 4, 57, 58, 60, 63, 80, 99
molecules, 14, 57, 58, 65, 75, 91
monoclonal antibody, 59
morphology, viii, 1, 19, 39, 61, 90, 91
mortality, 40
motor, 3, 7, 8, 9, 13, 15, 20, 21, 23, 30, 43, 45, 46, 47, 52, 56, 86, 87, 97, 106, 119, 122
motor neuron disease, 87, 97, 106
MPTP, 3, 12, 33, 55, 58, 59, 60, 64, 65, 66, 67, 68, 69, 70, 71, 72, 73, 87, 106
mRNA, 16, 20, 21, 30, 42
multiple sclerosis, 97, 119

mutation, 92, 94, 104, 105, 107
mutations, 101, 104, 105, 108, 112, 118, 121
myelin basic protein, 36
myelination, 17

N

nanovesicles, 54, 55, 57, 63, 70, 72
necrosis, 85, 88, 91, 114, 115, 121, 126
nerve, 50, 108
nervous system, viii, 2, 3, 4, 19, 22, 24, 25, 48, 50, 51, 52, 55, 57, 85, 111
neural development, 20
neurite, 18, 42, 65
neurobehavioral, 5, 11, 27, 30, 33, 36, 45, 50, 51
neurobiology, 38, 78
neuroblastoma, 60, 64, 65, 80
neurodegenerative, v, vii, viii, ix, 2, 4, 7, 26, 54, 55, 56, 58, 72, 83, 85, 87, 88, 90, 92, 95, 96, 98, 99, 101, 103, 104, 105, 108, 109, 110, 112, 113, 117, 119, 123, 124, 125, 126, 127
neurodegenerative diseases, vii, viii, ix, 2, 4, 7, 54, 55, 56, 58, 72, 83, 85, 87, 90, 92, 96, 101, 103, 104, 108, 109, 110, 112, 113, 117, 119, 124, 125, 126, 127
neurodegenerative disorders, ix, 86, 105, 108, 113, 126
neurodevelopment, v, vii, 1, 2, 4, 5, 14, 15, 16
neurogenesis, 2, 17, 19, 21, 24, 39, 42, 47, 52, 88, 109, 114, 120, 125
neuron, 7, 15, 87, 97, 98, 101, 106, 107, 112, 114, 122, 123, 125
neuronal death, ix, 85, 87, 88, 93, 94, 96, 99, 100, 102, 103, 105, 107, 108, 111, 114
neurons, 4, 10, 11, 13, 14, 15, 22, 23, 29, 31, 34, 35, 36, 38, 50, 54, 55, 57, 58, 72,

75, 82, 86, 88, 93, 97, 98, 103, 104, 105, 106, 107, 112, 113, 126
neurotoxicity, v, vii, viii, ix, 1, 2, 3, 5, 6, 7, 8, 9, 10, 11, 12, 13, 17, 19, 20, 21, 23, 24, 25, 26, 27, 28, 29, 30, 32, 34, 35, 37, 38, 39, 42, 43, 44, 45, 46, 47, 48, 49, 54, 65, 66, 71, 74, 75, 82, 85, 86, 87, 88, 93, 104, 105, 107, 108, 112, 115, 116, 118, 123, 125
neurotoxin, viii, 2, 59, 60, 66, 67, 71, 82, 85
neurotransmitter, 2, 11, 31, 51, 86, 94, 95, 99, 105, 111
neurotrophic, 10, 21, 29, 127
neurotrophic factors, 10, 21, 29
nigrostriatal, 7, 8, 13, 27, 105
nitric oxide, 12, 15, 20, 21, 23, 31, 36, 48, 87, 94, 95, 99, 110, 111, 113, 116, 122, 126
nitric oxide synthase, 12, 21, 31, 95, 116
noradrenergic, 10, 32
noradrenergic system, 10, 32
norepinephrine, 12
nutrient, 92, 97, 103

O

offspring, 5, 9, 11, 13, 15, 17, 18, 19, 20, 23, 33, 39, 40, 41, 42, 44, 47
oligodendrocytes, 17
oligomerization, ix, 54, 72
oligomers, 73, 79, 83
opportunities, 114
organelles, 13, 92
organic, vii, 1, 6, 14, 18, 22, 32, 35, 49, 51
organic solvents, 1, 6, 51
organophosphate, 9, 28, 29, 30, 31
oxidation, 10, 46, 99, 105
oxidative, ix, 2, 7, 9, 10, 11, 12, 13, 19, 21, 23, 24, 28, 29, 31, 33, 35, 39, 42, 43, 44, 46, 47, 49, 86, 87, 90, 91, 95, 99, 100, 104, 105, 107, 109, 111, 112, 120, 124, 125, 127
oxidative stress, 2, 7, 9, 10, 11, 12, 19, 21, 23, 28, 29, 31, 33, 39, 43, 44, 46, 47, 49, 86, 87, 90, 95, 99, 100, 104, 105, 107, 109, 111, 112, 120, 124, 125
oxygen, 96, 97, 99, 112

P

paints, 16, 23
paralysis, 107
paraquat, 12, 32, 33, 34
parathion, 10, 11, 29, 31, 32
passive avoidance, 11, 31
pathogenesis, ix, 4, 84, 109, 113, 114, 119
pathology, viii, 7, 54, 70, 73, 76, 77, 78, 86, 95, 98, 102, 105, 122, 125
pathway, 7, 15, 19, 36, 44, 48, 51, 57, 82, 84, 85, 89, 91, 100, 101, 103, 106, 108, 110, 116
PD, 3, 7, 12, 13, 54, 55, 56, 58, 59, 65, 66, 72, 73, 74, 85, 86, 87, 95, 97, 103, 105, 106
penicillin, 60
pentabromodiphenyl ether, 35, 36
peptide, viii, 54, 55, 65, 66, 80, 120
perinatal, 19, 39, 41, 42, 49
peripheral nervous system, 4, 22
peroxisome proliferator-activated receptors, 14
peroxynitrite, 96, 99, 122
pesticide, 7, 26, 30, 31, 33, 34
pesticides, vii, 1, 6, 7, 25, 26, 27, 29, 87, 127
pesticides, metals, alkenes, 1, 6
phagocytic cells, 90
phenotype, 26, 32, 60, 65, 126
phosphorylation, 13, 17, 21, 38, 46, 91, 102, 118
physical, 19, 86

physiology, 6, 20, 98, 99
plasma membrane, 56, 57, 58, 66, 72, 73, 74, 91
polyacrylamide, 47
polychlorinated biphenyl, 35, 36, 37
polychlorinated biphenyls, 3, 14, 35, 36, 37
polymerase, 12, 91, 115
population, viii, 1, 7, 86, 88, 89, 105, 110
postnatal, 2, 5, 8, 9, 11, 23, 27, 30, 31, 40, 47, 50, 124
postnatal days, 8
postsynaptic density protein-95, 18
preclinical, 7, 103
pregnant, 19
prenatal, 2, 5, 11, 17, 23, 25, 28, 30, 31, 33, 35, 36, 39, 43, 44, 49
preterm infants, 45
prevention, ix, 16, 24, 79, 108
primate, 106
progenitor cells, 17, 115
proinflammatory, 11
pro-inflammatory, 91
proliferation, ix, 12, 15, 23, 33
propagation, ix, 54, 58, 59, 72, 73, 74, 75
protein, vii, viii, 2, 5, 17, 19, 21, 22, 23, 36, 47, 54, 55, 57, 58, 62, 65, 66, 67, 70, 71, 72, 73, 74, 75, 76, 77, 78, 80, 81, 83, 86, 89, 91, 92, 97, 100, 101, 103, 104, 107, 108, 109, 110, 112, 113, 115, 116, 119, 120, 123, 124, 125
protein aggregation, 83, 86, 101, 110, 123
protein misfolding, 72, 112, 124
proteins, viii, ix, 7, 9, 16, 19, 21, 23, 38, 39, 40, 56, 57, 58, 71, 72, 73, 74, 79, 87, 89, 92, 97, 99, 100, 103, 107, 111, 113, 118, 122, 123
pups, 13, 14, 30
Purkinje, 9, 17, 22
pyrethroids, 8, 28

Q

quantification, 66, 67, 68
quinone, 115

R

radicals, 99, 106, 122
rats, 11, 15, 17, 19, 20, 21, 23, 24, 27, 28, 29, 31, 33, 36, 37, 38, 39, 40, 41, 42, 44, 46, 47, 48, 49, 50, 52, 110
reactions, 91, 94, 96, 98, 99
reactive oxygen, 13, 59, 95, 99, 112, 114, 123
reactive oxygen species, 13, 59, 95, 99, 112, 123
reactivity, 22, 96
receptor, 15, 16, 37, 38, 42, 47, 51, 78, 89, 91, 94, 100, 116, 118, 125, 126
receptors, 11, 14, 15, 16, 17, 18, 21, 24, 51, 80, 89, 91, 94, 97, 107, 118, 123, 126
recycling, 88
redox, 12, 99, 112, 113, 115, 122
reflexes, 11, 23
regeneration, 23, 50, 54
regenerative capacity, 100
relevance, 59, 81, 113
repair, 6, 102
replication, viii, 2
response, 2, 11, 13, 15, 36, 50, 65, 66, 72, 89, 91, 116, 119
retardation, 15
reticulum, 99, 100, 108, 111, 113, 116, 119, 120, 121, 122, 123
ribose, 12, 91, 115
RNA, 5, 17, 120, 123
room temperature, 62, 63
rotenone, 13, 34, 35, 87, 102, 106
routes, 6

S

secretion, 67, 70, 71, 72, 73
segregation, 73
sensitivity, 24, 34, 65, 120
sensitization, 107
serotonergic, 11, 32
serotonin, 12, 32
showing, 8, 21, 69, 72, 73
signal transduction, 51
signaling pathway, 9, 20, 41, 47, 86, 99, 102, 125
skeletal muscle, 22
social behavior, 7, 11
sodium, 3, 8, 18, 27, 41, 42
spatial learning, 11, 18, 24
species, ix, 13, 54, 59, 71, 72, 73, 74, 75, 95, 97, 99, 105, 112, 123
sporadic, 7, 79, 87, 104, 106
Sprague-Dawley rats, 31, 33, 42
stress, ix, 2, 7, 9, 10, 11, 12, 19, 21, 23, 28, 29, 31, 33, 39, 41, 42, 43, 44, 46, 47, 49, 86, 87, 89, 93, 95, 96, 98, 99, 101, 103, 104, 105, 107, 108, 109, 110, 111, 113, 114, 116, 118, 119, 120, 121, 122, 124, 125
striatum, 8, 12, 21, 33, 105
structural changes, 74
structure, 18, 52, 55, 74, 101, 113
subcellular, 13, 103
substantia nigra, 10, 12, 22, 97, 105, 113
substrate, 60, 91, 93, 116
susceptibility, 6, 27, 35, 47, 105, 106
susceptible, 4, 5, 19
synaptic, viii, 2, 12, 15, 16, 17, 18, 23, 24, 32, 33, 38, 40, 52, 79, 94, 108, 111, 118, 122, 124, 126
synaptic plasticity, 16, 17, 23, 24
synaptic transmission, 17, 33, 94
synaptogenesis, 14, 23, 38
synaptophysin, 16, 18, 23

T

T cells, 92
target, viii, 2, 6, 10, 27, 28, 32, 58, 90, 92
therapeutic targets, 102
therapeutics, 103, 108
thyroid, 2, 15, 36
thyroxine, 15
time periods, 4
tissue, 88, 90, 118, 121, 124
TNF-alpha, 89
TNF-α, 89
toluene, 23, 50, 51, 52
Toluene, 23, 51, 52
toxic products, 107
toxic substances, 40
toxicity, vii, 2, 3, 4, 5, 11, 12, 22, 26, 28, 33, 34, 37, 47, 48, 49, 67, 83, 86, 93, 94, 96, 119
toxicology, 27, 51, 117
toxin, 6, 55, 59, 62, 67, 68
trafficking, 9, 55, 58, 126
traits, 56, 81
transcription, 13, 101
transcriptomic, 9, 23, 28
transfer, viii, 19, 60, 62
transmission, ix, 69, 72, 79
transmission electron microscopy, 69
transport, 18, 21, 96, 97, 99, 100, 108, 122, 124
treatment, ix, 3, 13, 15, 17, 18, 20, 22, 50, 57, 60, 64, 65, 66, 67, 70, 71, 72, 73
triggers, 101, 110, 116
turnover, 12, 13, 73, 105
tyrosine hydroxylase, 8, 106

U

ubiquitin, ix, 7, 107, 110, 118
ubiquitin proteosome system, 101

unfolded protein response, 86, 100, 101, 116, 119

V

vesicle, 23, 71, 73, 92
visualization, 62
voltage-dependent anion channel 1, 55, 79

W

wastewater, 22
water, 6, 22, 40, 45
World Health Organization, 40
worldwide, 7, 18

Y

yeast, 92
yield, 12
young adults, 40

Z

zinc, 96, 99, 121

α

α-synuclein (α-syn), ix, 7, 12, 33, 34, 54, 71, 72, 78, 81

Related Nova Publications

FUNCTION AND METABOLISM OF AGING: LONGITUDINAL NEUROIMAGING EVALUATIONS

AUTHOR: Yongxia Zhou, PhD

SERIES: Neuroscience Research Progress

BOOK DESCRIPTION: The aim of this book is intended to provide both beginners and experts in biomedical imaging and health care a broad picture as well as new development in brain function and metabolism of aging using innovative neuroimaging techniques and advanced longitudinal /correlational analyses.

SOFTCOVER ISBN: 978-1-53615-613-3
RETAIL PRICE: $95

FACIAL EXPRESSIONS: RECOGNITION TECHNOLOGIES AND ANALYSIS

EDITORS: Flávia de Lima Osório and Mariana Fortunata Donadon

SERIES: Neuroscience Research Progress

BOOK DESCRIPTION: This book brings together contributions from different researchers on the theme of facial expressions, with an emphasis on emotional expressions, which may be of interest to professionals in neuroscience, technology and psychopathology.

SOFTCOVER ISBN: 978-1-53615-254-8
RETAIL PRICE: $95

To see a complete list of Nova publications, please visit our website at www.novapublishers.com

Related Nova Publications

CYTOCHROME C: ROLES AND THERAPEUTIC IMPLICATIONS

EDITOR: Natalia Arias

SERIES: Neuroscience Research Progress

BOOK DESCRIPTION: *Cytochrome c: Roles and Therapeutic Implications* provides a thoroughly revised, invaluable resource for university students and researchers in the life sciences, medicine and related fields.

HARDCOVER ISBN: 978-1-53614-907-4
RETAIL PRICE: $195

TEMPORAL LOBE EPILEPSY: PATHOLOGIC SUBSTRATES AND CAUSES

EDITOR: Richard A. Prayson, M.D.

SERIES: Neuroscience Research Progress

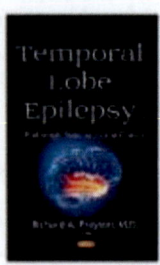

BOOK DESCRIPTION: The focus of this text is to review some of the underlying pathologic findings encountered in patients with medically intractable epilepsy who undergo the surgical resection of the epileptogenic focus.

SOFTCOVER ISBN: 978-1-53614-409-3
RETAIL PRICE: $82

To see a complete list of Nova publications, please visit our website at www.novapublishers.com